Hartford Steam Boiler Comp.

The Locomotive

Vol. XXXIII.

Hartford Steam Boiler Comp.

The Locomotive
Vol. XXXIII.

ISBN/EAN: 9783743376892

Manufactured in Europe, USA, Canada, Australia, Japa

Cover: Foto ©berggeist007 / pixelio.de

Manufactured and distributed by brebook publishing software (www.brebook.com)

Hartford Steam Boiler Comp.

The Locomotive

The Locomotive

OF

THE HARTFORD STEAM BOILER
INSPECTION AND INSURANCE CO.

VOL. XXXIII.

PUBLISHED BY
THE HARTFORD STEAM BOILER INSPECTION AND INSURANCE CO.

HARTFORD, CONN,
1920-1921.

INDEX TO VOL. XXXIII.—1920-1921

THE LOCOMOTIVE.—INDEX.

THE LOCOMOTIVE—INDEX.

The Locomotive

Devoted to Power Plant Protection

Published Quarterly

Vol. XXXIII.	HARTFORD, CONN., JANUARY, 1920.	No. 1.

EMPLOYEES WHO LEFT THIS COMPANY TO DEFEND
THE UNITED STATES AND ITS ALLIES AGAINST
GERMAN AUTOCRACY, INTRIGUE AND DOMINATION.
1914 — PERIOD OF WAR — 1919

UNITED STATES PERIOD IN THIS CONFLICT.
APRIL 6, 1917. JUNE 28, 1919.

ARMY	NAVY
DUDLEY P. ALLEN	W. H. ALTMANSBERGER
NORMAN R. ALLEN	ALBERT H. BAKER
CECIL L. BARRETT	THOMAS F. BARRY
EDWARD L. BENTLEY	ALDEN R. CHAMBERS
WILLIAM BYRT	LOUIS N. FAIRBANKS
WARD I. CORNELL	JAMES L. FOORD
WILLIAM CROSS	ARTHUR W. HIGHAM
HARRY D. DEAN	BENTON K. HUMPHREY
HENRY E. GERRISH	ARTHUR M. JOHNSON
ELMER B. RAINES	ALBERT H. MORRIS
GRAHAM B. HART	JOHN H. NOLAN
WILLIAM A. HARVEY	HENRY PONTET
JOHN M. LOVE	EUGENE RANG
JAMES H. LUCAS	LON J. REED
MICHAEL J. LYNCH	JOHN C. ROSS
WALTER L. MARLEY	JOHN SIMPSON
A. GORDON MERRY	GEORGE R. STICKNEY
EDWARD M. MURRAY	FREDERICK R. TAYLOR
WALTER S. PARKER	EDWARD A. TRUCKER
ARTHUR H. SEIFTS	EDWARD L. VAN WART
HASSEL A. WARREN	ARTHUR VIRGINIA
GEORGE A. WASSUNG	HUGO A. WEALING
	LAUREL M. WILLIAMS
	JOHN W. WILSON
	ALBERT R. WINTER
	SELINA L. WINTER
	JOHN H. WOLTERS, JR.
	HENRY J. VANDER ...

IN GRATEFUL RECOGNITION OF THEIR PATRIOTISM
THIS TABLET IS ERECTED BY
THE HARTFORD STEAM BOILER INSPECTION
AND INSURANCE COMPANY.

FIG. 1. VIEW OF GENERATOR

Explosion of a Large Steam Turbine.

THE accompanying illustrations are of parts of the wreck of a 3,000 K. W. turbo generator at the plant of the Connecticut Light & Power Company, Waterbury, Connecticut. This machine was destroyed by an accident which occurred on November 19, 1919, in which, most unfortunately, two men were killed and nine others more or less severely injured. The cause of the accident has not yet been definitely determined, we believe, but that it was of a violence and suddenness characteristic of an explosion is evidenced by the complete disruption of the machine, as our illustrations show.

The accident has been the subject of an investigation by the Connecticut Public Utilities Commission. According to a published abstract of the report of the Commission's engineer, Mr. A. E. Knowlton, the machine was in operation at 1:15 p. m. on November 19th, and apparently carrying its load in a normal manner. At about that time, the station lights dropped to a low brilliancy and then faded out. The report continues, "There was a definite time interval, not far from an actual minute, between the dimming of the lights and the wrecking of the turbine." This report goes on to say that the dimming of the lights was caused by the loss of excitation on two and perhaps all of the four generating units and that thus these units were suddenly relieved of load. Two of the turbines, after the accident, were found with their automatic speed limits tripped and the steam shut off. It would appear from the report also that the investigator was satisfied that there was no material increase in speed of the wrecked turbine in the interval between the dimming of the lights and its destruction. The evidence to substantiate this negative conclusion was not given in the published report. It, however, must have been convincing to the

FIG. 2. TURBINE END OF WRECKED MACHINE

investigator for, without definite evidence to the contrary, he would naturally attribute the wreck of a high speed apparatus, suddenly relieved of load, to the excessive forces of over speed.

The wreck of the turbine itself was so complete and its remains consist in the main of such small fragments that from them it was impossible to determine at the examination made by a representative of our company what part failed first. The frame of the generator and its bearing pedestals were broken up and even the bed plate was

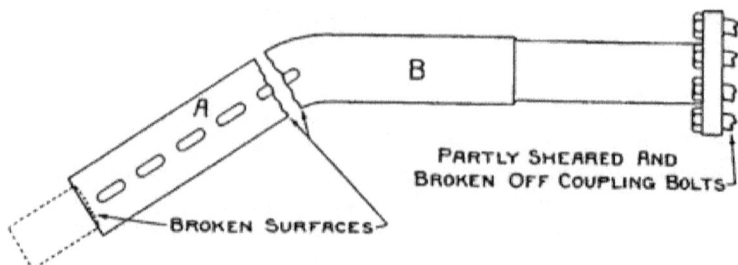

FIG. 3. SKETCH OF BROKEN TURBINE SHAFT

not spared. A peculiar and perhaps significant relic of the machine is the turbine shaft. This shaft was approximately 13" in diameter. It was broken in two places and the bolts of the flange coupling between the turbine and generator had partly sheared and broken off. One piece of this shaft, about 4 ft. in length, marked "A" in Fig. 3, after an extensive air flight, lodged in the basement under the turbine. Another piece, 8 ft. long, with the flange coupling, was hurled through a window into the yard. This latter piece, shown at "B" in Fig. 2, has a considerable offset, doubtless received an instant before the shorter piece of shaft was broken off. From the appearance of the bore of the turbine discs, all of which were considerably enlarged and belled out and from the fact that the piece of shaft on which they had been mounted was found after the wreck at some distance from the discs, there is reason to believe that the discs had slid off the shaft and started on their journey through roof and wall before the pieces of shaft commenced their respective flights. Of course, the blading in the turbine was broken up into small pieces. The rim of one of the discs, in which the blading was fitted, was completely ripped off.

Perhaps it was the character of the breaks in the turbine shaft in addition to other evidence against the over-speed theory which led the engineer of the Public Service Commission to the conclusion that the machine did not destroy itself by running away. As has been said, in such a complete break-up of all parts, as in this accident, one cannot determine where an initial failure occurred. The breaking of the very heavy shaft in two parts may, however, be significant and indicate that it first failed perhaps at the shouldered end shown in dotted lines in figure 3. If so, the portion of the shaft on which the discs were mounted would have had then one unsupported end and with the combined weight of the discs under such circumstances would have commenced to sag or bend. Only a slight amount of sagging or bending of the shaft would have caused the discs to rub and the blading to break up and at the great velocity at which it was running, the shaft, out of center at its free end, would begin to whip around with increasing radius until it took the sharp bend and broke off through the keyways of the innermost disc as shown in our diagram. In such a theory of the accident, the next step would have been the breaking of the inner turbine bearing and a repetition of the whipping around of the remaining piece of shaft until the coupling bolts gave way. Of course in any such theory of the accident there still remains to be explained the cause of the original break of the shaft at the point shown in the dotted lines of figure 3. It is difficult to attribute such failure to any shock received by a sudden release of load or from any other condition which the circumstances of the accident can supply. It perhaps seems to dodge the issue to suggest that fatigue of metal was the ultimate cause of the catastrophe, but if it is certain that over-speed did not occur, no other explanation seems possible. The machine had been in operation for several years during which, undoubtedly, it had carried its loads and over-loads without mishap and had been subject to and had successfully withstood the usual strains and shocks of public service operation. That after such service it was unable to bear the not very severe shock of a relief of load can only suggest that it had been weakened by the strains of those many years service.

We have ventured this rather long discussion of this turbine accident because of its unusual character even though we are unable to express a conclusion as to its cause. We trust our article may be suggestive of the unexpected dangers which threaten such high velocity, modern power producers. When anything goes wrong with such a machine, things happen in an exceedingly short space of time, usually without opportunity for human intervention to prevent the destruction or at least serious damage to costly pieces of property or to avoid death or injury of those in attendance. A turbine accident is, therefore, likely to be, as in the case we have written about, a catastrophe, and we feel that it is not improper for us to remind our readers that our company affords insurance against catastrophes of this kind.

On the Burning of Fuel Oil.

H. J. VanderEb.

THE shortage of coal and abnormal rise in coal prices of the last few years has given rise to a lively interest in the use of fuel oil under power boilers. Quite naturally, comparison as regards cost of coal and oil is the principal factor in this. For certain localities such a comparsion is at present favorable to the use of oil. Especially is this true for New England and other points on the Atlantic coast, remote from the coal fields. Add to this the uncertain delivery of coal of the present time from causes we need not here mention, and you have a fair index to the oil fuel situation.

As to how long these price relations can possibly continue, it is practically impossible to make a reasonably safe guess. From the present knowledge of the available world supplies of oil and coal which necessarily is rather vague, it seems however to be generally taken as a foregone conclusion that the oil supply will have ceased many centuries before coal will show signs of exhaustion. Undoubtedly at some future day, which may be in the lifetime of the present generation, the operation of the inexorable law of supply and demand may give back to coal the nearly undisputed monopoly it so long has enjoyed in the field of steam power generation. So long as the price remains favorable, however, oil will be a big factor in power plant operation and the present indications are that this may be for a number of years to come.

It is the purpose of this article to give a few helpful hints, gathered from the best information available, to steam users who desire to look into the desirability of changing from coal to oil for their boiler plants. In every case it is desirable that a reliable estimate be made in advance, of all the cost involved in making the necessary changes in the equipment. While the cost of the actual oil burning apparatus is light as compared with, for instance, mechanical stokers, there may be costly changes necessary in the boiler settings in order to obtain a reasonably high efficiency, which may drive the cost up to a disappointing figure. In addition to this, account should be taken of the possible necessity of installing extensive storage tank capacity depending on the proximity of the plant under consideration to an oil distributing center. For plants that are a considerable distance away from such an oil depot it is suggested to have a storage capacity of from thirty to forty days supply to take care of any interruptions of the regular delivery of the oil. Steam plants that have the good

fortune of being located right near an oil depot can of course avoid
a heavy investment in storage tanks and for such plants a week's
supply on hand might be considered sufficient. But even for such
installations, especially if they be public service plants with contracts
for their power output, business foresight may suggest the desirability
of the right proportion of reserve supply to insure continuous service
under unusual circumstances. With a further view to the possible
serious interruption of the oil supply at the source it has been suggested that no oil burning installation should be undertaken that
would not permit changing back to the use of coal in a reasonably
short time. In making estimates on proposed oil installations and
comparing the cost of the oil itself with that of coal, use can be made
of a handy approximate rule, sufficiently accurate for practical purposes, which is the simple relationship between the cost of the two
fuels as pointed out by Mr. W. M. McFarland. This is, that for
equal steam production the fuel cost will be the same when the number of dollars of the price of coal per ton (of 2,240 lbs.) is double
the number of cents of the price of oil per United States gallon.
This rule is based on the respective average heat values of oil and
coal per lbs. and takes into account the better efficiency obtainable
with oil than is possible with coal. Any other possible economies
incident to the use of oil such as the lower labor cost of handling fuel
oil as compared with that of coal and ashes are not included in this
rule.

The fuel oil that is at present sold for power purposes is, with
very little exception, the heavy residuum that remains after taking
off by partial distillation from the crude oil the valuable lighter
hydro-carbons, naphtha, gasolene and kerosene. This so-called "top-
ping" of the crude oil enhances the value of it as fuel rather than
diminishing it, as the flashpoint is thereby raised to a point where the
fuel can be handled with greater safety especially after being heated
to the temperature necessary for properly atomizing it at the burners.
The calorific value of the "topped" oil is not any less than that of
the crude oil, in fact it is even a little higher.

THE ADVANTAGES OF OIL OVER COAL

From a number of viewpoints oil is an attractive fuel for steam
generation. As already indicated in the foregoing it is possible to
obtain a higher efficiency with oil than with coal. It is comparatively
easier, so far as physical effort is concerned at least, to obtain almost perfect combustion with oil burning and keep out of the furnace

unnecessary excess air from the fact that there are no furnace doors opened every few minutes as with the hand firing of coal and there is no cleaning of fires with its attendant serious cooling off of the furnace. The required intensity of the heat from the burners is under practically instantaneous control to meet changes in the load. There is furthermore possible a considerable saving of labor in an oil burning plant as compared with that required for the handling of coal and ashes and there is a complete absence of dust.

For the small plant of one or two boilers a saving in the labor item should not be expected since of course for such an installation the same number of men will be required to tend to the burners as would be to shovel coal in the furnace. There are however many small plants where it could be expected of one man, with more justification from a safety standpoint, to tend to both the engine and boiler, if oil were used, than where he has considerable coal shoveling to do. But for the larger plant the labor economy is a real factor. One man can tend a considerable number of oil fired boilers with almost the same facility as he can one boiler. One other feature that may be mentioned in favor of oil fuel as compared with coal, is that the troubles of spontaneous combustion, so common with coal of high sulphur content, are entirely excluded with oil fuel.

There are almost no real disadvantages connected with oil burning to offset the several advantages mentioned. The one serious obstacle that can be mentioned is that in congested city districts the use of oil may be made prohibitive by local ordinances requiring special conditions with regard to location and isolation of storage tanks with a view to safety in case of fire. Some of this trouble however may be overcome by piping the oil underground to the plant from a point where oil can be conveniently stored with better safety.

The oil as received may contain a certain percentage of moisture which must be eliminated by giving it time to settle to the bottom of the storage tanks. It is therefore desirable to have always more than one tank for any conditions of required storage capacity so that the oil as it is used may always be pumped from a tank in which the settling of the moisture is as complete as practicable. Each tank should be provided with a bottom drain cock at its lowest point to run off any collected water or dirty oil.

At the ordinary outdoor temperatures, especially in the northern latitudes, it is necessary to heat the oil in the storage tanks to reduce its viscosity to a point where it can be pumped. As it is too wasteful to attempt to heat the whole tank to the desired temperature it

is entirely practicable to accomplish this by placing a steam coil right
near the point where the suction pipe enters the tank. On all piping
used for the transmission of oil it is desirable to have a steam connection so that they may be blown through and cleared of any
accumulations of silt which is more or less present in all fuel oil.
It is absolutely essential to have some effective form of strainer
placed in the suction pipe leading to the pump in order to catch the
fine grit and so to prevent undue wear of the pump cylinders. In
order to eliminate the pulsations of the pump, so that a steady flow
may be had at the burners, the pumps should be provided with an
ample air chamber.

Heating of the oil is furthermore a necessity to aid in the proper
atomization at the burner. It is most convenient and economical
with the heavy oils now being used to do the heating of the oil in
two steps, namely, to raise the temperature of the oil near the suction
outlet of the storage tank sufficiently to reduce its viscosity to a point
where it can be pumped and to have in the fireroom a separate heater
in which the oil can be given the desired temperature for proper
atomization. The final temperature of the oil just before atomization is usually about 140° to 160° Fahrenheit, where the oil is
atomized by means of steam, but it is best to find by trial the most
suitable temperature for each particular oil used to effect the best
economy. Great care should be exercised to not heat the oil above
its flashpoint. The flashpoint of an oil is the temperature at which
inflammable vapor begins to be liberated at its surface. Thermome-
ters should be present on the suction pipe leading from the storage
tank to the pump and on the pressure pipe between the pump and the
burners so that at all times proper control of the temperature may
be had. The inflammable vapors referred to are a distinct danger and
may give rise to disastrous explosions in the combustion space of
the boilers, when for instance the oil valve to a burner is inadvertently
left partly open under an idle boiler. Such gas explosions are known
to have done great damage to the setting walls and serious personal
injury. Aside from this danger the proper operation of the burners
is affected by the presence of vapor in the pipes as the oil will under
such conditions flow irregularly causing sputtering of the flame.

SOME DETAILS OF THE BURNERS

The function of an oil burner is to scatter the oil in a spray of
minute particles to make it possible for the oxygen of the air to
come in intimate contact with as much surface of the fuel as it is

feasible to expose to it. A solid stream of oil has a small surface as compared with the aggregate surface of all the minute oil drops that result when the solid stream is broken up into a fine spray. The work performed by the atomizing agent is simply the work of stretching the surface of the oil, hence the finer the spray the better are the chances for perfect combustion. The only limitation on the fineness of spray is the cost involved in producing it.

The burners that are most commonly used can be classified under two general types: 1st, spray burners in which the oil is atomized by means of a jet of steam or air, and 2nd, mechanical burners in which the oil is forced under considerable pressure through a small aperture of particular shape causing it to break up into small particles. As the small aperture of the mechanical burner will wear quickly larger by any grit in the oil, thus rendering it useless, the thorough straining of the oil is especially important when mechanical burners are used. It is however well for any type of burners to have a strainer in the pipe between the pump and the burners to catch any gritty or solid substance that may pass by the strainer in the pump-suction line.

Mechanical burners have an advantage over those that atomize the oil by means of an air jet because of the necessity of an air compressor with the latter type. They also have, theoretically at least, an advantage over steam spray burners because of the fact that all steam that is introduced into a furnace leaves the furnace (when combustion is complete) as steam, which carries with it some of the heat generated from the fuel, entailing a certain amount of loss. It is sometimes asserted that the burning of the hydrogen that is set free when the steam is decomposed by the high furnace temperature into hydrogen and oxygen, will add a certain amount of heating value to that of the fuel. The fallacy of this will be obvious when one considers that it takes just as much heat to decompose the steam into its component elements, hydrogen and oxygen, as can possibly be realized when these elements are again united by combustion.

Another advantage of mechanical burners over steam spray burners is that they are generally better adapted to take care of wider variations of load which necessarily is conducive to better economy under certain conditions of operation. However, the extreme simplicity of the steam atomizing burner and the excellent economy obtained with it when constructed on correct principles together with the comparatively low oil pressure and temperature it requires has made this type the favorite for stationary work. Burners using air as an atomizing agent are in successful operation but steam atomizing burners are used more generally. Wherever the loss of fresh water is not a vital factor the latter are usually the most satisfactory. The steam consumption has been found for the better make of burners to be approximately 2% of the total steam generated.

From a safety standpoint a so-called flat-flame burner is preferable over a burner producing a cone shaped flame for most types of boilers as it is simpler with the former to avoid the impinging of the flame on portions of the heating surfaces of the boiler. Localization of the intense heat of the flame on tubes or shell of a boiler will invariably result in overheating and blistering of the metal and should be carefully guarded against. Space forbids a detailed description here of the different types of burners on the market. Such of our readers as are interested in further perusal of this detail are referred to " The Science of Burning Fuel Oil " by W. N. Best and " Oil Fuel " by E. H. Peabody.

AMPLE COMBUSTION SPACE AN ABSOLUTE REQUIREMENT

The selection of the right type of burner, while of course important, is of less significance in obtaining the proper boiler efficiency than is the proper furnace volume and general design of the furnace. The ideal conditions of an oil furnace are that the particles of burning oil have an opportunity to linger just long enough in the furnace to be completely consumed before coming in touch with the relatively cool boiler surfaces which would extinguish them with the possibility that they are re-ignited higher up in the setting or in the uptake with a resultant waste of heat.

Ample space must therefore be provided in the primary combustion chamber; more indeed than for almost any other fuel. This extremely important fact may make the change from the use of coal to oil prohibitive for boilers that are set low.

It has proved feasible with existing coal burning boilers, in which the distance above the grate is not less than about 40 inches, to form a chamber for the oil flames by placing a layer of firebrick over the gratebars, leaving a sufficient number of openings in this layer of brick for air admission. It is safe to say however that it is best in any case, both from a safety and an economy standpoint, to remove the grate bars and install a flat checkerwork of firebrick to take the place of the grate, but placed close to the ashpit floor. leaving only sufficient space under the checkerwork to form an air duct.

The bridgewall should then be cut down to about the top of this checkerwork. In view of the high temperature to which the brickwork in an oil furnace is subject, which may reach 2800 degrees to over 3000 degrees Fahrenheit, only the best quality of firebrick should be used. It is impossible to give any sort of a definite rule for the proper amount of required combustion space. This can best be determined for each individual installation and its surrounding conditions by someone having extended experience with oil burning.

For water tube boilers of the inclined tube type the so-called "rearshot" burner should be used. This name applies to location of the burner rather than to type. It simply means that the burner is placed just in front of the bridgewall and shoots the flame toward the front of the boiler. The objects gained by this are that the flame projects in the direction in which the furnace increases in volume due to the fact that the tubes are inclined toward the front and the possibility of the flames impinging on the tube surfaces is practically excluded.

Fuel oil is successfully being used under vertical firetube boilers of the Manning type, but it is found that there is a tendency that not all the tubes participate in transmitting the products of combustion. The tubes directly over the burner proper are apparently idle while the tubes in the rear or the direction of the flame transmit all the heat. In such a case, good use may be made of retarders, consisting of spirally twisted strips of sheet metal, placed in the rear tubes which will have the effect of distributing the hot gases more uniformly.

FUEL ECONOMY HINGES LARGELY ON DAMPER ADJUSTMENT

The proper amount of draft through an oil burning furnace is a matter of great importance and on it hinges largely the success or failure of the installation in competition with coal. Less draft is required for the successful burning of oil than in the case of a coal furnace of the same relative capacity. The reasons for this are that with oil burning the draft does not have to overcome the same retarding influence as is produced by a fuel bed, and the action of the oil burner itself is moreover to some extent that of a forced draft. The volume of gases for a given rating is smaller with oil burning than with coal. From this it follows that it is not necessary to have as large a stack area for oil burning boilers, nor does the stack have to be as high as for coal burning of the same capacity. The proper amount of draft to be allowed when changing over an installation from coal to oil burning can be taken care of by keeping the stack damper partly closed, but it is better, and it makes the installation more fool-proof, to contract the area of the gas passage of a stack of too large capacity by means of a fixed plate with an opening in the center of the required size.

On the other hand there must be a sufficient draft suction to steadily carry off the products of combustion at a certain maximum rate which can only be determined by test for the best obtainable economy of fuel. If an insufficient amount of draft is allowed at the stack so that the action of the burner as a draft producer is relied on to push the gases, the action of the heat on the brickwork of setting walls and baffles will cause them to rapidly deteriorate. It is, therefore, a case of striking a happy medium between the evil of too much draft causing waste of fuel and that of not enough draft involving high upkeep cost.

As stated before the question of allowing just the right amount of draft is very important for proper economy and because of the fact that resistance to the draft is considerably less through an oil burning boiler than through a coal fired boiler, the handling of the damper is a much more sensitive operation with oil than it is with coal. It is, therefore almost needless to state that a suitable draft gauge, located so that it can be conveniently read, is practically indispensable when economy is desired. Carelessness in manipulating the draft will invariably lead to gross waste of fuel.

In one installation, that recently came to our notice, the records of oil consumption showed a " mysterious " gradual increase, until finally it was nearly double what it had been at first, although the steam output from the boilers was practically unchanged. The reason for this marked increase in the oil consumption was not far to seek. The emphatic and careful instructions, given at the time the oil equipment was installed, had " wore off " and the firemen had come to regard the close regulation of the draft as a useless bother. Consequently they were running with the stack damper and ashpit doors wide open causing a short white flame, which they no doubt regarded as hotter and therefore more efficient. The result was, as stated, a doubling up of the oil consumption. Here was a case where, with practically no effort, about ten barrels of oil per day could have been saved over a considerable period.

A clear stack on an oil burning installation is usually an indication that too much draft is passing through the furnace with a resultant low efficiency since all the unnecessary excess of air simply acts as a cooling agent and carries heat up the stack that ought to have served in making steam.

A slight haze coming from the stack indicates that conditions are more nearly ideal. In order to establish the best furnace conditions for any given load, the most satisfactory method is, of course, by means of flue gas analysis but in the absence of the proper apparatus for this, use can be made of a reasonably reliable and simple rule. When the furnace is well alight and the walls uniformly heated up to a high temperature, the draft should be pinched down by gradually closing the stack damper to a point where the flames have a slightly smoky fringe, when the damper should be opened again just sufficiently to clear the flames.

An Unusual Flywheel Accident.

FIG. 1

FLYWHEEL explosions are almost as a rule fraught with spectacular damage to the surrounding property and not infrequently with distressing loss of life and personal injury. It is rare indeed for a flywheel to go to pieces without heavy masses of metal being projected through roofs and building walls producing the effect of shell fire.

However, in the above illustration we show a case of genuine flywheel disruption due to centrifugal force, but there was no damage beyond some major fractures of the hub and arms of the wheel. The only usable picture we were able to secure of the wrecked flywheel does not clearly show that it was a complete loss. Such, however, was the case. The wheel in question was nine feet in diameter and was used in connection with a refrigerating engine. By the lucky coincidence that someone happened to be near the trottle and closed it at the moment the engine started to race, a complete disruption of the wheel was prevented. The wheel rim must have attained a considerable velocity as the links forming the rim-joints had stretched fully an inch due to the centrifugal force tending to separate the two halves of the wheel.

These links, which are round and of steel, are inside of the rim, a suitable recess being left in the rim to receive them, and they are secured to the rim sections by means of cotter keys. The only visible parts of the rim joint, therefore, after the wheel is assembled, are the ends of the cotter keys. They can be distinctly seen in the illustration, as can also the separation of the rim at the joint due to the stretch of the links.

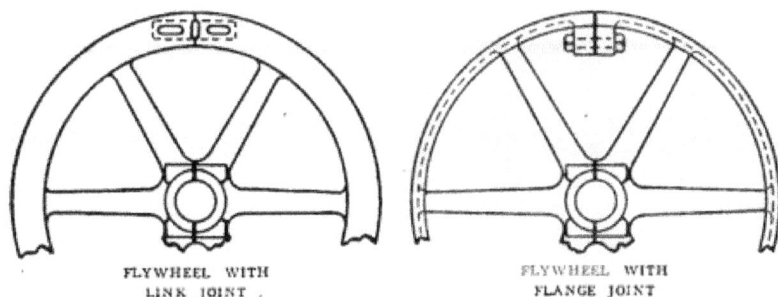

FLYWHEEL WITH
LINK JOINT .

FLYWHEEL WITH
FLANGE JOINT

FIG. 2

When the links commenced to stretch, such a severe strain was transmitted through the arms to the hub-bolts that two of them broke off and one section of the hub itself split in two, while one arm was completely broken free from the rest of the wheel.

While this is a good illustration how a flywheel of this type will explode if given sufficient speed, the case strikingly demonstrates the superiority of the link type of rim joint over the kind of joint formed by bolted flanges that are located in the rim midway between arms.

From the experience had with flywheel rims of the two types shown in Figure 2 it is safe to say that, had it been of the flange joint type, the rim would have disrupted before reaching anywhere near the velocity which produced the considerable stretching of the steel links.

The link type of rim joint when properly proportioned is the strongest possible joint. In calculations of allowable speed it is usually credited with 60% of the strength of a solid rim section. The reason why it is not used more extensively is that a comparatively heavy cross section of rim is required to apply it.

A flange type joint cannot be considered to possess a greater strength than 25% of that of the solid rim, while tests have shown that even this low percentage may not always be fully realized. Its greatest source of weakness is to be found in the fact that a bending action is produced in the rim by centrifugal force tending to throw the mass of metal constituting the bolted flanges out of the rim circle.

Explosion of an Auxiliary Steam Turbine.

AMONG the accidents reported to The Hartford Steam Boiler Inspection and Insurance Company was one of explosion of a small single stage steam turbine belonging to the Oliver Iron and Steel Company, Pittsburgh, Pa.

This little turbine stood in a sub-basement of the plant and was directly coupled to a centrifugal pump which was used for pumping water from the river. There had been some trouble with the water end of the unit due to fouling of the suction pipe, and at the time of the accident this was just thought to have been overcome and there were still a number of men standing near the little machine. The chief engineer, who was directing the starting up, stood near the water side of the machine and after it had been running for a short while' he signalled to the man at the throttle to shut off the steam. At about this moment the rotor burst and the flying fragments cut an opening through the casing all the way around its circumference. One man was killed outright and four others were badly scalded by the escaping steam. One of the latter, the chief engineer, died the following day from his injuries.

The cause of this accident has been attributed to derangement of the speed governor.

The Hartford Correspondence Course.

THE Firemen's Correspondence Course on combustion and boiler handling of the Hartford Steam Boiler Inspection and Insurance Company has taken its place as one of the established features of Hartford Service. The number of men now enrolled is well up into the hundreds, and some of the earlier students are finishing the course and receiving their certificates.

One thing, which is especially noticeable, with nearly every student is the marked improvement shown both in understanding and expression after the first two or three lessons. Apparently many of the men feel a bit awkward about the work at first and feel that they possibly may not be able to do it justice, but as they get into the swing of the first few lessons on combustion, which by the way is a subject of absorbing interest to most any one, they gain confidence and improve their work wonderfully. As soon as they have done, say, four or five lessons they become enthusiastic and the work not only improves but it comes in much more rapidly. This fact which has been very striking indeed has been a source of deep satisfaction to those who designed the course, and are responsible for its maintenance.

As an instance of how this course commends itself, we quote from the written opinion of an old and experienced fireman in the employ of one of our assured, a large corporation in the Southwest. This opinion was secured by our client to ascertain to what degree this method of study appealed to a man, who, from long experience, was an expert in firing matters. It was so satisfactory that it was circulated among the firemen of all our client's plants in an effort to interest the men to take the course. The employee whose opinion was asked. wrote the following:

"Will say in reply to your inquiry about Firemen's Educational Course from Hartford Boiler Insurance Company that it is O. K. I have not finished the course yet, but what few lessons I have received have helped me quite a lot. Although I have been a fireman for about fifteen years, I learn something new from each one of these lessons. I believe it a real good thing, and am glad that I took the course. Think it would benefit anyone who wishes to learn more about combustion."

Needless to say, we since have had the pleasure of enrolling a

THE H.

1866 - FIFTY THIRD YEAR OF THE HARTFORD - 1919

MAY

	Geo. H. Brown
1	Geo. H. Brown
2	E. H. Williams
3	S. B. Adams
4	C. B. Paddock
5	L. T. Greer
6	G. N. Delap
7	J. T. Coleman
8	R. T. Burwell
9	
10	
11	
12	
13	
14	
15	
16	
17	
18	

UPPER LEFT HAND CORNER OF VACATION SCHEDULE

considerable number of employes in the other power plants of this client of ours.

To anyone desiring to avail himself of this highly useful educational work we shall be pleased to mail descriptive circular and enrollment blank.

Vacation Schedule Blanks.

OUR Company has for several years past been publishing for the benefit of its policyholders a convenient blank form for recording vacation periods. It has been our practice in various years to mail these out to all of our clients. The circumstances of many of our policyholders probably prevent their use of this blank to advantage, while to many in turn they are of such convenience that their omission now would be a matter of regret. We have felt that this year we should not waste paper and expense in the indiscriminate distribution of these blanks. We will publish them as usual and furnish a supply to each of our department offices and general agents listed on the back of THE LOCOMOTIVE. In those departments they will be available for forwarding to such of our policyholders as have found them useful and wish the blanks for the vacation season of 1920. Accordingly, those who want them should apply to the nearest office or general agent of the company.

The cut on opposite page will indicate to those who are unfamiliar with this blank how it may be used. At the left the names of officers or employees entitled to a vacation are listed in column. The chart is divided into months, weeks and days by vertical lines. The vacation is recorded by a dot or cross for each day of the period on the line with the individual's name. When the record is complete, not only is the vacation clearly shown recorded but also at a glance it may be determined how many employes will be absent at any given time.

"The graspin'est man I ever knowed," said Uncle Jerry Peebles, "was an old chap named Snoopins. Somebody told him once that when he breathed he took in oxygen and gave out carbonic acid gas. He spent a whole day tryin' to find out which of them two gases cost the most if you had to buy 'em. He wanted to know whether he was makin' or losin' money when he breathed."—*Safety Hints.*

The Locomotive

Devoted to Power Plant Protection
Published Quarterly

HARTFORD, JANUARY, 1920.

Single copies can be obtained free by calling at any of the company's agencies.
Subscription price 50 cents per year when mailed from this office.
Recent bound volumes one dollar each. Earlier ones two dollars.
Reprinting matter from this paper is permitted if credited to
The Locomotive of the Hartford Steam Boiler I. & I. Co.

THE two turbine explosions recorded in this issue of THE LOCO-
MOTIVE will be of absorbing interest to many who in the
past have had misgivings about the claims that have been put
forward at times that steam turbines can be made proof against explosion. A number of conspicuous turbine accidents have occurred
in the last few years and, as their number in use becomes greater,
failures of large turbo-generators as well as of the smaller types of
turbines used for auxiliaries may be more frequently heard from.

The larger one of the two turbines whose failure is described in
this issue has the unique distinction of being the first large steam turbine, so far as we know, to completely run to destruction beyond the
possibility of any of its parts being salvaged and it would seem in this
instance that this is to be laid to the breaking of the shaft.

This feature of shaft breakage would appear to be deserving of
more than passing notice. The stresses in a turbine shaft are of two
kinds, namely, there is a torsion stress and a bending stress, the former
being a steady stress, but the latter is one of complete reversal from
a compressive stress to one of tension during every revolution of the
shaft and it is likely to be accentuated by any vibration present, however slight.

For both of these stresses the usual design of turbine shaft has a
very high factor of safety. However, from the research work done
with varying and repeated stresses we know that the ultimate failure

of the material may be brought about by apparently harmless loads if such reversing stresses are repeated a number of billions of times.

It appears barely possible that the billions of repetitions of the reversal of stress that take place in certain portions of a turbine shaft during its continuous operation over a time period of say a decade needs the serious attention of engineers responsible for the safe operation of this type of prime mover.

After the United States entered the world war over 10% of all the employes of this Company entered active service in the Army or Navy of this Country. In addition at least two enlisted in the service of the Allies.

The war is over, many of the employes have returned to the Company, some are at this writing still in the service, and just two will never return. Lieutenant John M. Love was killed in the Argonne Forest while leading his company, his Captain having been killed, and Yeoman Thomas Barry died on shipboard on what was to be his last trip before his retirement.

The officers of this Company, appreciating the patriotism of these employes, and in recognition of the event, have caused to be erected a bronze honor tablet at the Home Office of the Company.

With a desire to further honor these men and one woman, we have the pleasure of illustrating on the front page of this issue of THE LOCOMOTIVE a picture of the tablet. May all respect be paid to these people for their service to their Country.

Personal.

It is with regret that we announce the resignation of Mr. C. C. Perry, who, since July 1st, 1912, has been the Editor of THE LOCOMOTIVE.

Mr. Perry accepted a position in the Engineering Department of the Ætna Casualty & Surety Company of Hartford and entered on his duties there on December 1st, 1919. Mr. Perry, by education and experience, had a broad acquaintance with engineering matters, and his ability as a writer on technical subjects is shown by the issues of THE LOCOMOTIVE which have appeared during the last six years. We congratulate the Ætna Company in securing his services, and wish him the best success in his new position.

Mr. C. R. Summers has been appointed Chief Inspector of the Atlanta Department. For several years past Mr. Summers, as Assistant Chief Inspector, has directed the inspection work of the Atlanta Department with excellent results. During the fourteen years of his connection with the Company he has built up an enviable reputation as a thorough mechanic and boiler expert among our clients in the South. His new appointment comes as a well-earned promotion.

Mr. L. T. Gregg has been appointed Chief Inspector of our Cleveland Department. Mr. Gregg came into the service of The Hartford Steam Boiler Inspection and Insurance Company in 1911, serving a number of years as an inspector. In the last three years the active part of directing the inspection work of the Cleveland Department has been largely done by Mr. Gregg as Assistant Chief Inspector. His long experience with the Company, as well as his previous training, eminently fit him for his new responsibilities.

One other appointment was occasioned by the recent establishment of a branch office at New Orleans. The large amount of work and the rapid expansion of business of our New Orleans Department made necessary some assistance in the supervision of the inspection work. To fill that need, Mr. Eugene Unsworth was appointed Assistant Chief Inspector. Mr. Unsworth has had a long experience as an inspector, having been in the employ of The Hartford Steam Boiler Inspection and Insurance Company since 1906.

A Correction.

In spite of the careful editing and proofreading of all matter entering in THE LOCOMOTIVE, an error has slipped by in the October issue, on page 233 at the top.

In the equation there stated the symbols X and Y were erroneously used for a and b.

The equation should have read:

$$b = \frac{\tfrac{1}{2} \times 1\tfrac{1}{4} \times 4\tfrac{3}{8} + 2 \times \tfrac{1}{2} \times 1 + 3 \times \tfrac{1}{2} \times \tfrac{1}{4}}{\tfrac{1}{2} \times 1\tfrac{1}{4} + 2 \times \tfrac{1}{2} + 3 \times \tfrac{1}{2}}$$

$b = 1.315$ inch, and $a = 5 - 1.315 = 3.685$ inch

so that it may correspond with the symbols in Fig. 3 on page 232 and the text just above the equation. To those who preserve the article in question for future reference we suggest that they make this correction.

Flywheel Explosions, 1919.

(Continued from April 1919 Locomotive, page 185.)

(10.) — A shaft governor located in a flywheel disrupted due to a fulcrum pin shearing off, at the plant of the American Linoleum Company, Lenoleum, Staten Island, N. Y. on March 19th. The damage to the engine on which the wheel was mounted was considerable.

(11.) — On April 21st a rope drive failed at the plant of the Ashgrove Lime and Portland Cement Company, Chanute, Kan., causing the breaking up of two 7 foot wheels.

(12.) — On April 23rd an 8 ft. driven wheel and one 22 ft. driving wheel of a rope drive exploded at the Steel Works of the Interstate Iron and Steel Company, Marion, Ohio, doing great damage to the mill.

(13.) — On April 28th a 6 ft. wheel on a gas engine exploded at the planing mill of Elias Bros., Buffalo, N. Y.

(14.) — A flywheel on a buzz-saw exploded on May 16th at the Johnson farm about two miles distant from Avon, N. Y. Fragments of the wheel were thrown a quarter of a mile. One man was severely injured.

(15.) — On May 22nd a flywheel exploded in the plant of the Consumers Company, Chicago, Ill. One man was killed.

(16.) — A 14 ft. flywheel exploded on June 15. at the plant of the City Electric Light Company, Douglas, Ga., due to a run-away when the governor belt broke. The damage was several thousand dollars.

(17.) — On July 10th a 16 ft. flywheel exploded at the plant of the International Rubber Cloth Company, West Barrington, R. I. Four men were injured. The damage was $7,700.00.

(18.) — A large pulley on a dynamo burst on July 11th at the refrigerating plant of the J. W. Hester Company, Savannah, Ga. One man was mortally injured.

(19.) — The governor balance weight located in a flywheel broke off damaging the wheel beyond repair at the plant of the Dodge Steel Pulley Corporation, Oneida, N. Y. on August 19th.

(20.) — On August 29th an engine wheel failed at the plant of the Broadway Coal Mining Company, Simmons, Ky.

(21.) — A 12 ft. flywheel exploded on Sept. 2nd at the Racine Industrial Works, Racine, Wis.

(22.) — A 13 ft. flywheel exploded on Sept. 3rd in the 10" Chill Mill of the Pittsburgh Steel Company, Glassport, Pa. Two men were injured. The damage was $17,000.00.

(23.) — A 54" pulley failed on Sept. 11th at the plant of the Clayville Paper Mills Company, Clayville, Oneida County, N. Y.

(24.) — On September 15th a gas engine wheel exploded at the Municipal Power Plant, Norwood, Ohio. A fragment of the wheel broke a gas main. The fire that followed did damage estimated at $50,000.00.

(25.) — A 6 ft. flywheel on a variable speed engine exploded on Sept. 27th at the plant of the Ironsides Board Company, Norwich, Conn.

(26.) — On October 6th a 3 ft. pulley failed at the plant of the Northern Paper Mills, Greenbay, Wis.

(27.) — On October 13th a flywheel burst at the plant of the Pacific Coast

Coal Company, New Castle, Washington. One man was killed by escaping steam from a steam main broken by a fragment of the wheel. One other man was seriously injured.

(28.) — A 12 ft. wheel burst on October 13th at the plant of the Imperial Sugar Company, at Sugarland, Texas.

(29.) — A large gear failed on October 17th at the plant of the Apsley Rubber Company, Hudson, Mass.

(30.) — On October 24th a 20 ft. wheel exploded in the Puddle Mill of the Penn. Steel and Iron Works, Lancaster, Pa. One man was killed and 6 men were severely scalded by escaping steam and otherwise injured by flying debris. One section of the exploded wheel went up through the roof and upon returning to earth glanced the side of a boiler cutting a hole in the shell 3" wide by 14" long.

(31.) — On November 9th a flywheel burst on the Wolf Spring Oil Lease, near Allentown, N. Y. Two men were instantly killed.

(32.) — On Nov. 19th a steam turbine exploded at the plant of the Connecticut Power and Light Company, Waterbury, Conn. (For detailed account of this accident see page 2 of this issue.)

(33.) — On December 2nd a small turbine exploded at the plant of the Oliver Iron & Steel Company, Pittsburgh, Pa. (For detailed account see page 16 of this issue.)

(34.) — On December 7th a flywheel left the engine of a sawmill of Mr. Tom Young, near Ambrose, Ga., when at full speed. One man was killed.

Boiler Explosions.

MARCH, 1919.

(99.) — One section cracked in a cast iron sectional heating boiler on March 3rd at the Elizabeth's Industrial School, New York, N. Y.

(100.) — A 4-inch nipple pulled out of the drum on a water tube boiler doing damage to brickwork on March 1st, at the veneer and lumber mill of C. B. Willey, Chicago, Ill.

(101.) — A boiler exploded on March 3rd in the shop of the Syracuse Corner Block Co., Syracuse, N. Y.

(102.) — On March 3rd the blow-off pipe pulled out of elbow at the Kissel Motor Car Company's plant at Hartford, Wis., resulting in the severe injuries of one of the firemen.

(103.) — When starting an engine on March 3rd the cast iron steam separator ruptured due to water hammer at the Bergman Knitting Mills, Philadelphia, Pa.

(104.) — A boiler ruptured on March 3rd at the main plant of the State Penitentiary, Joliet, Ill.

(105.) — A tube ruptured March 4th in a water tube boiler at the plant of the Ebensburg Coal Co., Colver, Cambria County, Pa. Two persons were scalded, one seriously.

(106.) — On March 5th a tube ruptured in a water tube boiler at the plant of the Scovill Mfg. Co., Waterbury, Conn.

(107.) — Three sections cracked on March 6th in the cast iron sectional boiler belonging to J. W. Welch, Omaha, Neb.

(108.) — A section cracked March 6th in a cast iron sectional heating boiler at the Bellevue Hotel, Denver, Colo.

(109.) — An air receiver exploded on March 9th at the Rendering Plant of Swift and Company, Harrison, N. J.

(110.) — Five sections cracked on March 10th in a cast iron sectional boiler belonging to R. H. Gardiner and Philip Dexter, Trustees, due to low water.

(111.) — On March 10th a rupture occurred in the flange turn of a drum head of a water tube boiler at the plant of the Buckeye Cotton Oil Co.

(112.) — Five sections cracked in a cast iron sectional boiler belonging to W. T. Duker Co., Quincy, Ill., on March 10th.

(113.) — Three sections cracked on March 10th in a cast iron sectional boiler of Louis L. Friedman, Perth Amboy, N. J.

(114.) — A furnace flue collapsed in a marine type boiler on March 13th in the basement of office building belonging to Henry L. Corbett et al., Portland, Oregon, causing fatal injuries to the engineer and two of his assistants.

(115.) — Five sections cracked in a cast iron sectional boiler on March 15th at the Magaziner Baking Corporation, Springfield, Mass.

(116.) — A rendering vessel exploded on March 13th at the abattoir of Charles Maybaum & Son, Newark, N. J. The building in which it was contained was completely wrecked.

(117.) — A heating boiler exploded on March 16th in the basement of the Hoiles Block, Alliance, Ohio.

(118.) — While cutting in a boiler the blind end of a steam header was blown off by water hammer action on March 16th at the plant of Greenwood Cotton Mills, Greenwood, S. C.

(119.) — A tube ruptured in a water tube boiler on March 16th in the plant of the Chicago Coated Board Co., Chicago, Ill.

(120.) — On March 17th a tube failed in a Hawley Furnace under a boiler of the Central Ice & Cold Storage Co., New Orleans, La. Two men were scalded, one fatally.

(121.) — A tube ruptured on March 19th in a water tube boiler of the Cohankus Mfg. Co., Paducah, Ky.

(122.) — A tube ruptured in a water tube boiler on March 18th at the Blue Grass Plant of the International Agricultural Corporation, Mount Pleasant, Tenn., injuring and scalding two men.

(123.) — A drier exploded on March 19th at the plant of Mulsen, Klein and Krouse Mfg. Co., St. Louis, Mo.

(124.) — On March 19th a steam pipe burst at the plant of the L. Candee Rubber Co., New Haven, Conn.

(125.) — On March 20th a tube ruptured in a vertical water tube boiler at the plant of the Fletcher Paper Co.

(126.) — On March 20th, a blow-off let go under a boiler at the Rhode Island Institute for the Deaf, Providence, R. I., scalding two of the attendants.

(127.) — A tube ruptured in a water tube boiler on March 20th at the Public Service Corporation at the plant at Passaic River & Coal Sts., Newark, N. J., causing slight injuries to one man.

(128.) — A heating boiler exploded on March 20th in the garage of L. C. Muss, Weatherly, Pa., blowing the roof from the building.

(129.) — A boiler of a freight locomotive of the D. L. & W. Railroad blew up on March 22nd, a mile west of Cresco. Three men were instantly killed.

(130.) — On March 22nd, a cast iron head of a paper drying cylinder blew up, doing considerable damage to the paper mill of the Hammermill Paper Co., Erie, Pa.

(131.) — A boiler blew up on March 22nd, at the Yokohama Laundry, San Mateo, Cal. The Japanese proprietor and a laundry worker were instantly killed.

(132.) — On March 22nd a 20 H. P. power boiler operating a sawmill outfit blew up in Sharpsburg Township, North Carolina. The boiler belonged to Lewis Scott, who was running the mill but escaped injury. One other man was badly scalded.

(133.) — Ten sections cracked in a cast iron sectional heating boiler on March 25th in the basement of the Beaver County Court House, Beaver, Pa.

(134.) — On March 26th a tube ruptured in a water tube boiler belonging to the American Gas & Electric Co., at Wheeling, W. Va. One man was injured.

(135.) — When cutting in a boiler which had been out for repairs a sizable piece blew out of the stop valve body at the main boiler house of the Dare Lumber Co., Elizabeth City, N. C. on March 26th. Two men were severely scalded.

(136.) — On March 29th a cast iron hot water heating boiler exploded at the House of the Good Shepherd, Hartford, Conn.

(137.) — A section cracked in a cast iron sectional boiler on March 29th at the garage of the Standard Tire & Auto Co., New Britain, Conn.

(138.) — Three sections cracked in a cast iron boiler on March 29th at the Home Club, Richmond, Va. Six sections cracked in the boiler next to it at the same time.

(139.) — Two sections cracked on March 30th in a cast iron sectional boiler at the garage of William Krauss Garage Co., New York City, New York.

(140.) — A tube ruptured on March 29th in a water tube boiler at the plant of the Industrial Iron Works, Bay City, Mich.

(141.) — The fire sheet ruptured due to low water on March 30th in a boiler belonging to the St. Paul Coal Company, Cherry, Ill.

(142.) — A safety valve fitting was blown apart March 31st on a small boiler at the Star Confectionery Company, Boston, Mass.

(143.) — One cast iron header cracked on March 30th on a water tube boiler of the American Bridge Company, Elmira Heights, N. Y.

(144.) — Two tubes burst in a water tube boiler on March 31st in the plant of the Water & Light Department, Knightstown, Ind.

APRIL, 1919.

(145) — A blow-off pipe ruptured and pulled out of a fitting on April 2nd at the plant of the North American Chemical Company, South Milwaukee, Wis.

(146) — A large tank used in the making of dyes exploded on April 2nd at the plant of the California Ink Company, Berkeley, Cal. It crashed through

the second floor and the roof of the building and after an air flight fell down through the roof of an adjoining building. The damage was very large.

(147) — A boiler ruptured on April 2nd at the plant of H. Murphy and Son, Pittsburgh, Ohio.

(148) — On April 3rd an ammonia tank exploded at the plant of the Rieck-McJunkin Dairy Company's plant, New Castle, Pa. Of the three men that were working on the tank when the accident occurred, one was almost instantly killed and two were very seriously injured.

(149) — A boiler exploded on board the steamer "Cape Breton" while four miles out from St. John, Newfoundland on April 6th. The Chief Engineer with two Chinese firemen were killed while two other men were seriously injured.

(150) — A section in a cast iron boiler cracked on April 5th at the plant of the Snowflake Axle Grease Company, Fitchburgh, Mass.

(151) — A cast iron boiler cracked on April 5th at the Garfield School, Bridgeport, Conn.

(152) — A section cracked in a cast iron sectional boiler on April 7th at the Library Building, Alliance, Neb.

(153) — On April 8th a boiler exploded at the plant of the Sinclair Oil Refining Company, Buffalo, N. Y. The damage was very large.

(154) — A kier in the dye house of the Joslyn Mfg. Company, Olneyville, R. I. exploded on April 7th, one man was very severely injured, and the damage to the property was very large.

(155) — A shell of a drier failed on April 10th due to wear at the plant of the Point Milling & Mfg. Company, Mineral Point, Mo.

(156) — Three sections cracked on April 10th in a cast iron sectional boiler at the Baxter Hotel, Idagrove, Iowa.

(157) — Three cast iron headers of a water tube boiler failed on April 11th at Nueces Hotel, Corpus Christi, Tex.

(158) — An autogenously welded air tank exploded on April 11th at the plant of the York Mfg. Company, York, Pa.

(159) — The shell plate of a boiler failed on April 12th at the plant of the Clawson Chemical Company, Ridgeway, Penn.

(160) — A cast iron header of a water tube boiler cracked open on April 12th at the plant of the Rochester & Lake Ontario Water Company, Rochester, N. Y.

(161) — A boiler exploded on April 13th at the Locust Gap Plant of the Philadelphia & Reading Coal & Iron Company. Two men were killed.

(162) — A boiler ruptured on April 13th at the plant of the Hurni Packing Company, Sioux City, Ia.

(163) — A tube ruptured on April 16th in a water tube boiler at the Buckingham Ave. Power Station, of the Public Service Corporation of N. J. Perth Amboy, N. J.

(164) — On April 16th a crown sheet collapsed in a locomotive boiler of the Kirby Lumber Company, Houston, Tex.

(165) — A shell ruptured of a boiler of the Boonville Creamery & Cold Storage Company, Boonville, N. Y. on April 17th.

(166) — On April 18th a boiler explosion occurred on the U. S. S. Beaukelsijk. Two enlisted men were killed.

(167) — A boiler exploded in a Turkish Bath House, New Haven, Conn., on April 17th; two men were injured.

(168) — A boiler exploded on April 19th at the plant of Otto Zovler & Co., Chicago, Ill. One man was killed and five others injured.

(169) — The blow-off pipe failed on a boiler at the Lufkin Ice Company, Lufkin, Tex., on April 20th.

(170) — Two brace rivets snapped off their heads and were blown out of the tube sheet of a water tube boiler on April 20th at the plant of the Hess & Hopkins Leather Co., Rockford, Ill.

(171) — A tube ruptured in a water tube boiler on April 20th at the plant of the United Electric Company, Lexington, Neb.

(172) — A tube burst on April 20th in a water tube boiler belonging to the Pacific Gas & Electric Company, Fresno, Cal. One man was severely scalded.

(173) — A tube ruptured on April 22nd in a water tube boiler of the Columbia Chemical Company, Barberton, Ohio.

(174) — A hot water boiler exploded in the basement of the Anhilde Apartments, New York City, on April 22nd.

(175) — A tube burst in a water tube boiler on April 25th at the plant of the Timkin Detroit Axle Company, Detroit, Michigan.

(176) — A section cracked on April 27th in a cast iron sectional boiler of the Holyoke Supply Company, Holyoke, Mass.

(177) — A sizable piece blew out of a tube on April 29th in a water tube boiler of the D. C. Shepard Strong & Warner Company, St. Paul, Minn.

(178) — A tube ruptured on April 30th in a water tube boiler of the Florida Ice & Coal Company, Jacksonville, Fla.

MAY, 1919.

(179) — A tube burst in a water tube boiler of the Public Service Corporation of Perth Amboy, N. J., on May 4th.

(180) — A vertical cast iron sectional boiler cracked on May 4th at the Pinehurst Hotel, Laurel, Miss.

(181) — A tube ruptured in a water tube boiler at the Perth Amboy plant of the Public Service Corporation of New Jersey on May 6th.

(182) — A hot water boiler burst in the basement of the McKinley Apartments, Patterson, N. J. on May 9th. The two children of the janitor of the building were slightly injured.

(183) — While standing on a siding the boiler of a locomotive blew up of the Seaboard Air Line R. R., Raleigh, N. C. Three men were instantly Killed, May 13th.

(184) — On May 14th a blow-off pipe of a pulp digester of the Atlantic Paper & Pulp Company, Savannah, Ga., broke between the vessel and the gate valve allowing the contents to escape. Three men were injured, one of them seriously.

(185) — On May 14th a steel elbow of a blow-off pipe ruptured on a boiler of the American Candy Company, Milwaukee, Wis.

(186) — A tube burst and four cast iron headers fractured on May 14th in a water tube boiler of the Illinois Northern Utilities Company, Freeport, Ill.

(187) — On May 14th a cast iron header fractured in a water tube boiler at the Dodge Mfg. Company, Mishawaka, Ind.

(188) — A tube burst on May 14th in a water tube boiler at the plant of the Emmerling Brewing Company, Johnstown, Pa. One man was injured.

(189) — On May 15th a tube burst causing the cracking of two cast iron headers in a water tube boiler belonging to Swift & Company, Union Stock Yards, Chicago, Ill.

(190) — On May 19th a tube burst in one of the water tube boilers of the Noblesville Heat, Light & Power Company, Noblesville, Ind. One man was severely injured.

(191) — On May 22nd a boiler of a locomotive belonging to the Pennsylvania R. R. blew up near East Palestine, Ohio. Two men were instantly killed and another was severely injured.

(192) — A boiler blew up on May 22nd at the plant of the Humble Ice & Power Company, Humble, Tex. One man was killed, and one was severely injured.

(193) — On May 23rd an air tank exploded at the plant of the Central States Granite Company, Huntington, W. Va. One man was so severely injured that he died a week later.

(194) — On May 23rd a rupture occurred due to bulging out of the top of drum shell of a water tube boiler of the Nebraska Cement Company, Superior, Neb.

(195) — A tube burst on May 25th in a water tube boiler of the Nichols Cooper Company, Newtown, L. I., N. Y.

(196) — On May 30th a tube ruptured in a water tube boiler at the plant of McKinney Steel Company, Josephine, Penn.

(197) — On May 29th while cutting in a boiler water hammer action occurred blowing off the top of an 8" non-return stop valve at the Bleachery Power Plant of the Lonsdale Company, Providence, R. I. One man was scalded to death.

(198) — A furnace collapsed on May 31st in a boiler belonging to Truax's Criterion Laundry, Omaha, Neb.

The Hartford Steam Boiler Inspection and Insurance Company.

ABSTRACT OF STATEMENT, JANUARY 1, 1919.

Capital Stock, . . . $2,000,000.00.

ASSETS.

Cash in offices and banks	$361,295.49
Real estate	90,000.00
Mortgage and collateral loans	1,505,900.00
Bonds and stocks	5,121,486.85
Premiums in course of collection	654,112.42
Interest accrued	108,152.83
Total assets	$7,840,947.59

LIABILITIES.

Reserve for unearned premiums		$3,429,363.68
Reserve for losses		153,378.80
Reserve for taxes and other contingencies . .		367,147.68
Capital stock	$2,000,000.00	
Surplus over all liabilities	1,891,057.43	
Surplus to Policy-holders, ...		**$3,891,057.43**
Total liabilities		$7,840,947.59

CHARLES S. BLAKE, President.

FRANCIS B. ALLEN, Vice-President,   W. R. C. CORSON, Secretary.

L. F. MIDDLEBROOK, Assistant Secretary.

E. S. BERRY, Assistant Secretary and Counsel.

S. F. JETER, Chief Engineer.

H. E. DART, Supt. Engineering Dept.

F. M. FITCH, Auditor.

J. J. GRAHAM, Supt. of Agencies.

BOARD OF DIRECTORS.

Incorporated 1866.

Charter Perpetual.

INSURES AGAINST LOSS OF PROPERTY

AS WELL AS DAMAGE RESULTING FROM

LOSS OF LIFE AND PERSONAL INJURIES DUE TO EXPLOSIONS OF STEAM BOILERS OR FLY WHEELS

Department.	Representatives.
ATLANTA, Ga.,	W. M. FRANCIS, Manager.
1103-1106 Atlanta Trust Bldg.	C. R. SUMMERS, Chief Inspector.
BALTIMORE, Md.,	LAWFORD & McKIM, General Agents.
13-14-15 Abell Bldg.	R. E. MUNRO, Chief Inspector.
BOSTON, Mass.,	WARD I. CORNELL, Manager.
4 Liberty Sq., Cor. Water St.	CHARLES D. NOYES, Chief Inspector.
BRIDGEPORT, Ct.,	W. G. LINEBURGH & SON, General Agents.
404-405 City Savings Bank Bldg.	E. MASON PARRY, Chief Inspector.
CHICAGO, Ill.,	J. F. CRISWELL, Manager.
160 West Jackson St.	P. M. MURRAY, Ass't Manager.
	J. P. MORRISON, Chief Inspector.
	J. T. COLEMAN, Assistant Chief Inspector.
CINCINNATI, Ohio,	W. E. GLEASON, Manager.
First National Bank Bldg.	WALTER GERNER, Chief Inspector.
CLEVELAND, Ohio,	H. A. BAUMHART, Manager.
Leader Bldg.	L. T. GREGG, Chief Inspector.
DENVER, Colo.,	J. H. CHESNUTT,
918-920 Gas & Electric Bldg.	Manager and Chief Inspector.
HARTFORD, Conn.,	F. H. WILLIAMS, JR., General Agent.
56 Prospect St.	E. MASON PARRY, Chief Inspector.
NEW ORLEANS, La.,	R. T. BURWELL, Mgr. and Chief Inspector.
308 Canal Bank Bldg.	E. UNSWORTH, Ass't Chief Inspector.
NEW YORK, N. Y.,	C. C. GARDINER, Manager.
100 William St.	JOSEPH H. McNEILL, Chief Inspector.
	A. E. BONNET, Assistant Chief Inspector.
PHILADELPHIA, Pa.,	A. S. WICKHAM, Manager.
142 South Fourth St.	WM. J. FARRAN, Consulting Engineer.
	S. B. ADAMS, Chief Inspector.
PITTSBURGH, Pa.,	GEO. S. REYNOLDS, Manager.
1807-8-9-10 Arrott Bldg.	J. A. SNYDER, Chief Inspector.
PORTLAND, Ore.,	McCARGAR, BATES & LIVELY,
306 Yeon Bldg.	General Agents.
	C. B. PADDOCK, Chief Inspector.
SAN FRANCISCO, Cal.,	H. R. MANN & Co., General Agents.
339-341 Sansome St.	J. B. WARNER, Chief Inspector.
ST. LOUIS, Mo.,	C. D. ASHCROFT, Manager.
319 North Fourth St.	EUGENE WEBB, Chief Inspector.
TORONTO, Canada,	H. N. ROBERTS, President Boiler Inspection
Continental Life Bldg.	and Insurance Company of Canada.

The Locomotive

DEVOTED TO POWER PLANT PROTECTION

PUBLISHED QUARTERLY

Vol. XXXIII.	HARTFORD, CONN., APRIL, 1920.	No. 2.

Flywheel Explosion at Scotia, Cal.

DAMAGE TO ENGINE HOUSE BUILT OF 12" CONCRETE WALLS.

THE illustrations on the front cover and next page of the present issue of the LOCOMOTIVE show some of the resulting damage when a 16 foot flywheel exploded on February 23rd at the sawmill of the Pacific Lumber Company, Scotia, Cal. The property damage was quite heavy but fortunately there were no fatalities as is so often the case with such disasters. Of the five men who were only slightly injured, four did not even stop working.

The wheel was of the flange joint type, made in four sections and really consisted of two wheels bolted together which formed a total rim face of 5 feet. The engine on which the exploded wheel was mounted was of the twin type with releasing type of valve gears, both cylinders receiving steam at approximately boiler pressure. One fly-ball governor regulated the cut-off for both cylinders with the usual arrangement of a rockershaft extending from one side of the engine to the other and this feature appears to have had a more or less prominent part in the cause of the engine running away.

The leading cause of the accident was a sudden release of all load from the engine by the breaking of the main driving belt which is thought to have been brought about by a piece of wood wedging between the belt and the face of the wheel. The broken main belt knocked off the governor-driving belt from its pulley so that the governor stopped revolving. If the trouble had terminated there the dire results shown in the illustrations should have been averted by the so-called safety cams on the valve gear which come into play when the governor drops to its lowest position and prevent the further admission of steam to the cylinders. However, as the big belt coiled up into the wheel pit, it broke the afore-mentioned governor rockershaft, which

GENERAL VIEW OF CYLINDERS AND VALVE-GEAR.

had the effect of preventing the safety cams on the north cylinder from functioning so that this cylinder continued to receive steam during a portion of each stroke and the speed of the engine increased rapidly.

We are advised that when the trouble began one of the operators ran to the throttle valve wheel and succeeded in closing the valve to within ⅛ inch after which he ran for safety just in time to escape with his life. This throttle valve was located overhead in the steam-line at the point where the steam-line forked into the two pipes that led to the cylinders, and moreover, on each cylinder stood a spacious steam separator so that the volume of steam between the throttle and the cylinders was very large and in all probability was sufficient to run the engine at a destructive speed, so that it is extremely doubtful whether his courage would have been crowned with success even if he had shut the valve tight.

The loss was estimated between twenty-five and thirty thousand dollars and was covered by insurance.

Horizontal Tubular Boiler Settings and Details of Installation.

H. E. Dart, Superintendent of Engineering Department.

OUR Engineering Department is now engaged in making new drawings of setting plans for horizontal tubular boilers. In past years there has been a big demand for such setting plans and some of the tracings for the more common sizes of boilers are literally worn out. In making the new drawings, advantage is taken of the opportunity to show certain features in greater detail than was formerly the case and the scope of the plans has also been extended so as to include typical methods of piping and the proper manner for installing the usual fittings and attachments. Figures are also given to show the quantities of bricks required for setting the boilers in accordance with the plans. For each of the common sizes of boilers, it is the intention to make four drawings, two with overhanging fronts and two with flush fronts, one of each style showing boilers suspended independently of the setting walls and the other showing boilers supported by means of brackets resting on the walls. The complete set of plans is not yet finished but drawings are ready for many of the ordinary sizes of boilers and blueprints can be furnished from such drawings as are finished. Requests for such blueprints should be made preferably through the chief inspector of the department in which the boilers are located (see list of departments on back of THE LOCOMOTIVE) rather than directly to the Engineering Department, because our chief inspectors are familiar with the conditions which exist and are generally able to submit the data which we need to determine which drawing is best adapted to each particular case.

The most important features in connection with the new setting plans are described below. While many of the features mentioned will apply equally well to the design of settings for any other type of boiler, it should be remembered that this description is concerned primarily with hand-fired horizontal tubular boilers using coal for fuel, and is written from that viewpoint.

WALL CONSTRUCTION.

On our old setting plans the outside walls are shown as indicated by Type I, Figure A, but on the new plans we are showing all of the four types of construction described in Figure A. leaving it to the boiler owner to make his choice between these designs. Complete dimensions are given on the drawings for each type of construction.

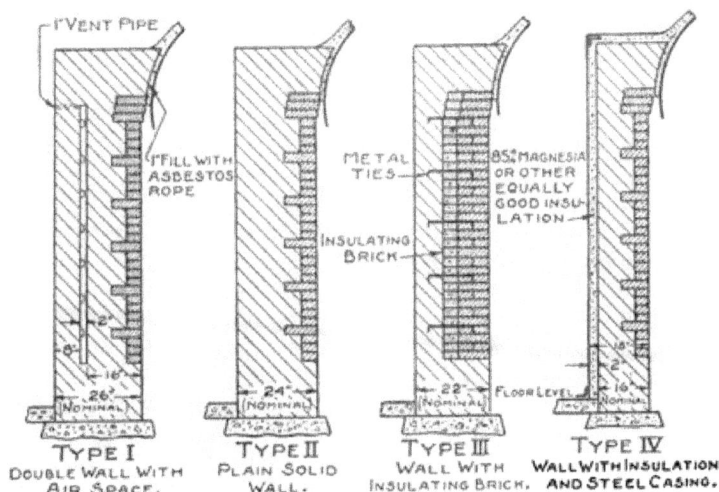

Fig. A.   Different Types of Construction for Setting Walls.

The design shown by Type I involves the construction of two separate brick walls, bonded solidly together for a distance of about sixteen inches at the top and at the bottom, but separated by an air space two inches wide for the remainder of the height.   It is thought by many people that this air space acts as a heat insulator but such is not the case; experiments by the Bureau of Mines have shown that a wall of this type will transmit just as much heat under given conditions as a solid wall of the same total thickness.   As regards air leakage into the furnace, however, the double wall with air space has a distinct advantage over the solid wall shown by Type II because the cracks will occur principally in the inner wall, leaving the outer wall intact.   With a solid wall, the cracks will extend clear through the brickwork, thus greatly increasing the probability of air leaks and thereby decreasing the efficiency on account of excess air.   Not long ago we had occasion to make an examination of a solid-wall setting which had been built in the same boiler room with two older settings of the air space type. Although the new setting had been in use only a few months the test with a candle flame showed more leaks than were found in the other settings which had been used several years. Of course such a test is not entirely conclusive, since there are other features which should be considered, but we believe it gives a fair indication as to what may be expected in the average case. In constructing setting walls with an air space it is advisable to insert a few ventpipes as indicated in the cut, these pipes being especially desirable if the bricks contain much moisture when they are laid. After the setting has thoroughly dried out, all ventpipes should be permanently sealed so as to prevent air leakage into the setting and heat radiation from the inner wall.

Type III in Figure A makes use of insulating bricks to reduce the amount of heat that is transmitted through the wall and thereby lost. These insulating bricks are made of different materials by different manufacturers and they are cut to the proper size to lay up evenly with ordinary bricks and fire bricks. They have little mechanical strength in themselves so that it is best to use metal ties, as shown in the cut, for bonding the inner firebrick section to the common brick on the outside. It is also advisable to use a uniform thickness of nine inches for the firebrick lining in place of the 4½ inch lining with headers as shown for the other types. This type of construction makes a very good setting, costing somewhat more than either Type I or Type II.

Type IV is similiar to Type I with a steel casing substituted for the outer wall and the air space filled with magnesia or other good insulating material. This makes a most excellent form of setting, the only drawback to its more general use being its greater cost as compared with other types. The insulating material reduces the heat radiation loss to a minimum and the steel casing prevents the even greater loss due to air leakage through the setting walls. Furthermore, a setting of this kind presents a very neat appearance and requires less space than any of the other types illustrated, there being a saving of eight inches in length and sixteen inches in width as compared with Type I. Number 8 U. S. gage steel plates should be used for the casing with angle irons placed about 3½ feet apart along the sides and back and with similar angles at the top, bottom, corners and elsewhere as needed for stiffness and stability.

For the division walls between boilers set in battery the style of construction shown in Figure B is satisfactory, regardless of what type of construction is used for the outside walls. The vertical slot shown in the center of the wall does not indicate an air space like that in Type I but is intended to show that the two walls should be built separately and not bonded together in the center. This is advisable to make allowance for expansion when there is a fire on only one side of the wall.

The sections in Figure A apply to the side walls at the rear of the bridge wall. For the furnace section in front of the bridge wall, we advise that the walls be battered from the grate level to the closing-in line near the middle of the boiler shell. Our drawings show a batter of six inches in this height, thus making the walls that much thicker at

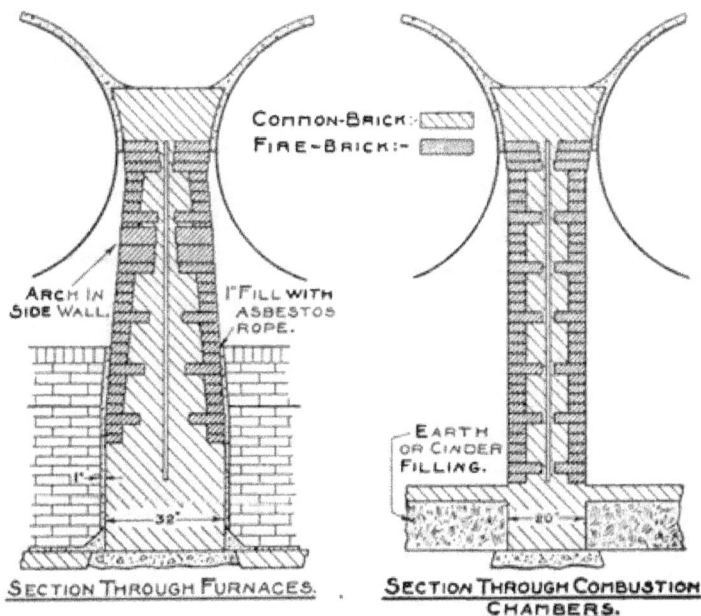

Common-Brick:-
Fire-Brick:-

Arch in
Side Wall.

1" Fill with
Asbestos
Rope.

Earth
or Cinder
Filling.

1"

32"

20"

Section Through Furnaces.   Section Through Combustion
                                             Chambers.

Fig. B. Division Walls Between Boilers Set in Battery.

the bottom.   A reference to Figure B will make this point clear.   The section at the left shows the battered wall while that at the right shows the straight form which can be used back of the bridge wall. This figure shows sections for the division wall between boilers but the same idea should be applied to the outside walls.

In constructing side walls and division walls it is a good idea to build an arch in the firebrick lining at a height of about three feet

If escape pipe is used it should have a diameter at least equal to that of the safety valve to which it is attached and if it is over six feet long it should be supported independently of the safety valve with such arrangement that no strain will come on the valve body.

Connection for draft gage.

Sampling tube for flue-gas analysis.

Gate Valve ordinarily left wide open.

Check Valve.

Union.

Globe Valve with boiler pressure on top of disk.

Asbestos rope.

Damper with provision for hand operation from floor.

Floor level.

Two valves of the outside-screw-and-yoke type should be used if boiler is connected to a common steam main with one or more other boilers. Valve nearest boiler should be preferably of the non-return type.

Escape Pipe should be drained by a drip ell or other suitable means with opening at least ¾in diameter.

Stop Valve should be placed as near the boiler as practicable.

Arch in side wall.

¾" Pipe for draft gage.

Steam Pipe should drain away from valve.

Downward Pitch

Set rear end of boiler 1 inch lower than front end.

2" or 2½" Blow Off Pipe, extra heavy pipe and fittings.

V-Shaped Pier of fire brick.

Drain required if two valves are used. Drain pipe should have open end visible while operating main valves.

Steel Elbow

Center of Fusible Plug 2" above line of upper surface of tubes.

Cover this trench with steel plate or loose brick.

Blow Off Pipe should have two valves, or valve and cock if pressure exceeds 125 lbs. (100 lbs. in Massachusetts)

Clearance of ¾" all around pipe inside of sleeve. Space filled in with asbestos to prevent air leakage.

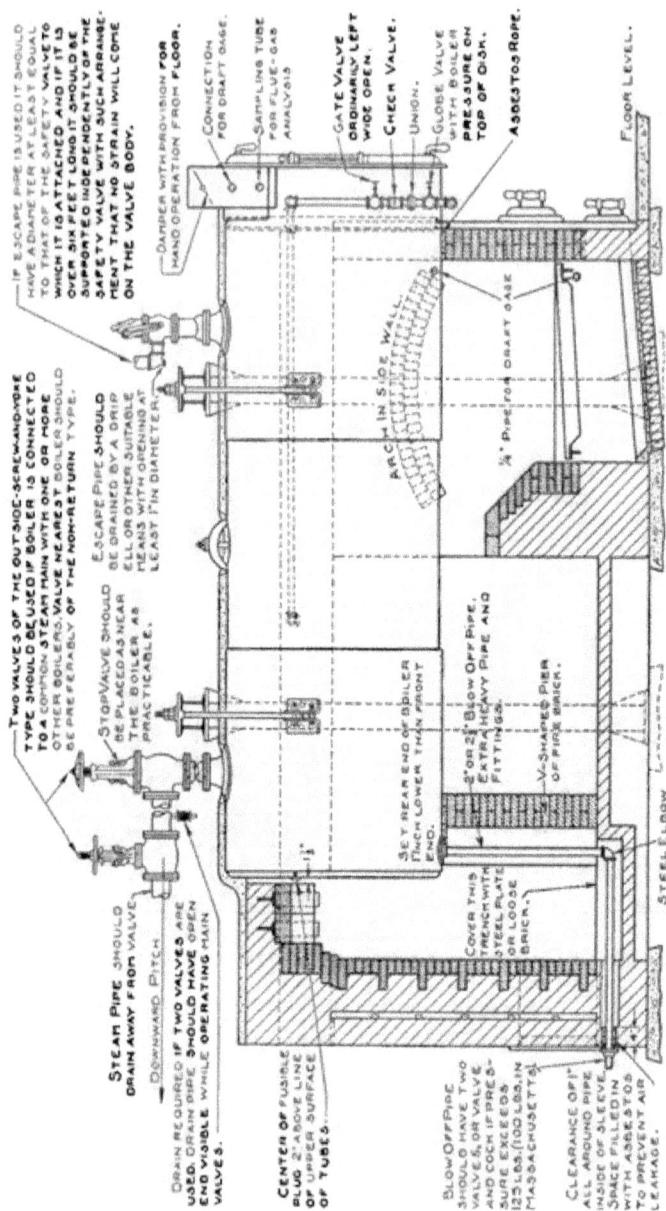

Fig. C. Longitudinal Section Through Center of Setting.

above the grates, as illustrated in Figure C. When it is necessary to replace firebrick this arch supports the brickwork above and prevents it from falling down.

For construction like that shown in Types I, II, and IV, where the firebrick lining is only 4½ inches thick, headers should be used for every fifth course or even more frequently. In all firebrick work the joints between the bricks should be made just as thin as possible. For this reason a trowel should not be used but the bricks should be dipped in thin fire-clay and then rubbed down into place so as to make "brick-to-brick" joints.

ALLOWANCE FOR EXPANSION AND PREVENTION OF AIR LEAKS.

Ample provisions should be made throughout to allow the boiler and the setting to expand without cracking the brickwork or opening up places where air can leak into the setting. If the brickwork is built tight up to the boiler shell at the closing-in line, cracks are sure to develop when the boiler is heated up and there will also be an opportunity for air to leak in between the boiler and the brickwork. It is best, therefore, to leave the brickwork about an inch away from the boiler and fill this space with asbestos rope or some similar material, as illustrated in the different sections of Figure A. In a similar way, the brickwork and the ironwork of the boiler front should be kept about ¾ inch away from the boiler shell (and concentric therewith) and this space should be filled in with asbestos rope. To prevent cracking due to endwise expansion of the bridge wall, it should be built separately from the side walls, leaving a space of about one inch at each end. This space should be filled with asbestos rope to prevent the accumulation of ashes which would become solid and nullify the advantage to be gained by building the bridge wall independently of the side walls.

At the rear end of the boiler, a space of about 1½ inches should be left between the boiler head and the brickwork; this space can best be sealed against air leakage by extending the insulating covering down over it as shown in Figure C. There is a tendency for the covering to crack open at this point as the boiler expands and contracts but this difficulty can be largely overcome by the use of a piece of sheet iron, formed to fit over the rear end of the boiler shell and bent down over the head to extend out on top of the brickwork. With the piece of sheet iron in place under the covering the probability of cracking is lessened and, if a crack does develop, the sheet iron will tend to prevent air from leaking into the setting. We advise the use of insulating

covering for the boiler top instead of the brick arches which are sometimes used. The covering is a better heat insulator and it can be
removed and replaced more readily in case repairs or inspections of
the boiler shell are required. The covering can best be applied in the
form of blocks which can be held in position with mesh wire and then
finished with plastic magnesia or other insulating cement to make a
smooth finish and fill the joints between the blocks. A harder surface
can be secured by using a little Portland cement in the final coat. The
total thickness of the insulation should be at least two inches.

The loss due to excess air is generally greater than that from any
other cause in hand-fired boilers of the type under consideration, and
it is also the most difficult to prevent because it is such an intangible
sort of thing that the firemen cannot be made to realize its importance.
Much of this excess air leaks in through the setting walls and efforts
to prevent such leakage by the methods outlined above will be well
repaid. It is not sufficient merely to construct the setting as described,
however; inspections and tests should be made at frequent intervals
to be sure that the asbestos remains in place and that the joints are
properly sealed at all points. We have made a number of investigations of this kind and we almost invariably discover air-leaks at some
of the places mentioned above as well as around blow-off pipes, firing
doors, clean-out doors, etc.

There are several different paints and coatings on the market
which can be used to good advantage in the prevention of air leaks.
Such compounds are usually composed of asphalt, asbestos and other
materials, combined to produce a thick elastic coating which will
stretch without cracking as the setting expands when it is heated up.
The coating is usually applied to the entire surface of the setting, care
being taken to work it into all cracks, joints and openings around door-
frames, boiler fronts, or other similar places. A very satisfactory
home-made substitute can be prepared to take the place of the commercial compounds.

PROTECTION OF BLOW-OFF PIPES.

The proper protection of the blow-off pipe is an important feature
in connection with any setting for a horizontal tubular boiler. There
are many ways of providing such protection and in making a choice
between different methods one important principle to be kept in mind
is that the pipe should be easily accessible for inspection. For this
reason a simple pipe sleeve around the blow-off pipe is not satisfactory
because such a sleeve cannot be removed without disconnecting the
blow-off pipe. Split sleeves of cast iron are better but it is usually

rather difficult to remove them after the connecting bolts have been exposed to the heat and flames. Several patented styles of blow-off covering are available and these give good results as a rule. In general, such blow-off coverings are made of some refractory material and applied in sections with an interlocking arrangement so that they are easily removable.

Except under extraordinary conditions, the method of installation shown in Figure C provides ample protection for blow-off pipes. The principal features of this method are a V-shaped pier of firebrick which prevents the flames from impinging upon the vertical portion of the pipe, and the location of the elbow in a covered trench where it will be well protected. Blow-off pipes are more liable to fail at the elbow than at any other point and the location of the elbow in this position is therefore highly desirable. The best arrangement is to build the bottom of the combustion chamber at a somewhat higher level than the boiler-room floor so that there will be space enough to install the blow-off valve or cock without cutting into the floor. It is advisable also to locate the cleaning door at one side of the center where there will be no interference with the blow-off valve when the door is opened. Plenty of space to permit freedom of movement, due to expansion or settlement, should be left around the pipe where it passes through the setting wall. For this purpose a pipe sleeve about four inches long should be built into the brickwork at the outer end but a larger opening can be left around the pipe through the remainder of the wall thickness, without any sleeve. The sleeve should have a diameter two inches greater than that of the blow-off pipe and it should be filled with asbestos to prevent air leakage. A set-screwed collar on the pipe makes a good finish against the brickwork together with provision for a gasket of sheet asbestos or other suitable material to more thoroughly seal the opening against air leakage. The V-wall should be left a little below the boiler shell to allow for expansion and settlement, and the space should be filled with asbestos to keep the flames from impinging upon the flange where the pipe is connected to the boiler. Blow-off valves should always be located so that there will be ample opportunity for a man to get away in case of a break in the blow-off piping while he is operating the valves.

ARCH-BARS.

The rear arch-bars shown on our setting plans and in Figure C are of the so-called " HARTFORD " type, designed by this Company several years ago. Bars of this type extend transversely of the setting,

spanning its width and bearing upon the side walls. Except for large
boilers, only two of these arch-bars are needed for a single setting
but a different pattern is required for each size of boiler. In some
sections of the country the "quadrant" type of arch-bar is more
popular and it is just as acceptable; this style of arch-bar is made in
the form of a quadrant or ninety-degree arc of a circle. The bars rest
on the rear wall, arching over to the rear head, and some means must
be provided to support the upper ends, so as to permit the boiler to
expand without developing air leaks. Several of these bars are needed
for a single setting, the exact number depending upon the diameter of
the boiler, but the same pattern can be used for all sizes of boilers
where the distance from the rear head to the rear wall of the setting
is the same. Both types of arch-bar described above are so designed
that the metal is protected by the firebrick and not exposed to the
action of the flames and hot gases; this feature should be a require-
ment in the design of any arch-bar.

Arch-bars should be set so as to leave a full, free opening through
all the tubes, with proper provisions for inspecting and removing the
fusible plug but, at the same time, care should be taken that no part
of the head above the lowest permissible water level is exposed to the
heat. We recently heard of a case where a head was burned, due
either to poorly designed arch-bars or to placing the arch-bars so high
as to expose the upper part of the head to extreme heat.

GRATES.

We believe that there is a general tendency to use larger grates
than necessary with hand-fired boilers of the horizontal tubular type
and this belief is borne out by our experience in several cases where
we have found that coal was being burned at a rate of ten to twelve
pounds per square foot of grate area per hour whereas better results
would be obtained with a rate of fifteen to twenty pounds of coal per
square foot per hour. In some cases we have advised blanking off the
rear part of the grates by covering them with fire brick and a gain in
economy of coal consumption has been secured in such cases. Furthermore, it has been common practice to use the same size of grates for
a given diameter of boiler regardless of the tube length though it
is obvious that if a certain area is proper for a boiler eighteen feet long
it would not be correct for a sixteen-foot boiler in which the heating
surface would be about eleven per cent less. Assuming an evaporation
of about nine pounds of water per pound of coal, the ratio of heating
surface to grate area should be about 40 to 1 in order to develop the
full rated capacity of a boiler when burning coal at the rate of fifteen

pounds per square foot of grate per **hour**. In designing our new setting plans we have used this ratio to determine the grate area, within the limits imposed by commercial standards as regards length of grates. Provision **for** overloads and allowance for a lower rate of evaporation can be taken care of by burning the coal at as high a rate as twenty pounds per square foot of grate per hour, this rate being attainable with proper draft and good firing methods. On the other hand, many horizontal tubular boilers in small plants are never operated at their **full** capacity and in such cases the grate area could be even **smaller**. It **is** fully realized that a larger grate area may be desirable in certain special cases but it is believed that the ratio of 40 to 1 will give good results for the average case and, of course, this is all that a set of **general** plans **can be expected to cover**; special cases should be considered in the light of **all** the data available in each instance.

With battered furnace walls, **as** described in the foregoing, the width of grates **will be** six inches less **than** the boiler diameter while with straight walls, as frequently used, **the** grates have a width equal to **the diameter of** the boiler. As explained above, the grate area is **generally larger than it** should be and **the** smaller dimension for width **is therefore generally** satisfactory. **For** the sake of simplicity and **uniformity, stationary** grates are shown **on all** of our setting plans but **we recommend the use** of shaking grates under ordinary conditions.

HEIGHT ABOVE GRATES.

Remarkable savings in **fuel** consumption **have been claimed** in many cases as a result of setting horizontal **tubular boilers at extreme** heights above the grates but, as a general **rule, these claims do not** seem to be fully substantiated because all of the credit for any increased efficiency is laid to the greater height of furnace whereas there are usually other factors which should also receive consideration. In a typical case of this kind, a new boiler is installed in a plant where there are one or more older boilers and perhaps the new boiler is set at a height of 5 feet above the grates while the corresponding height of the old boilers is only 28 inches. More or less careful tests are made and it is found that the new boiler is more economical in coal consumption than the old ones. It is then almost invariably assumed that the gain in economy is entirely due to setting the boiler at a greater height above the grates; although the old settings may be twelve or fifteen years old and full of cracks and openings which permit the entrance of a large percentage of excess air while the new setting is tight, this fact is completely disregarded. Furthermore, it seems to be generally assumed that the height chosen in any such case is the proper height to give the best results although there is usually no information available to prove that just as good results would not have been obtained with a height of 3½ feet, for instance, instead of 5 feet. In attributing a gain in economy to higher settings there are other factors also which may be ignored such as a change in the fuel used, an improvement in the proportions or design of the new boiler as compared with the older ones, better firing methods, improved draft conditions and method of draft control, a better type of grates, etc.

For any set of fixed conditions as regards size of boiler, character of fuel and other details, it is evident that there must be some limit in height of furnace beyond which there will be no gain in economy. It would probably not be possible to fix such a limit very definitely but much interesting information could be obtained from a series of carefully conducted tests carried out by some agency such as the Bureau of Mines which would have the necessary apparatus and technical skill, together with the means for insuring that all other conditions remain constant while the height of the furnace is varied.

The combustion volume, and therefore the height from grates to boiler shell, should be varied in accordance with the character of the fuel used, more volume being required for fuels containing a large proportion of volatile matter than for those which contain a relatively greater percentage of fixed carbon. On our setting plans we do not show any fixed dimension for the furnace height but we recommend certain dimensions as determined from our experience and best judgment. For a 72-inch boiler with tubes 18 feet long, for instance, the heights which we advise would be as follows:—

For anthracite coal and semi-bituminous coals containing less than 18 per cent. of volatile matter (Pocahontas, Georges Creek, etc.)— 36 inches.

For bituminous coals containing from 18 per cent. to 35 per cent. of volatile matter (Pittsburgh)— 40 inches.

For bituminous coals containing more than 35 per cent. of volatile matter (Illinois, etc.)— 44 inches.

For other sizes of boilers the figures are varied so as to maintain approximately the same ratio of combustion volume to grate area. In this connection it might be mentioned that the ratio of combustion volume to grate area would be nearly 19 to 1 for a 72-inch boiler with 18-foot tubes, set as shown in Figure C and with a height of 44 inches from grates to boiler shell. Although the setting is designed only for hand-fired horizontal tubular boilers, this ratio is considerably in excess of that ordinarily used for stoker-fired water tube boilers which may be forced to 100 per cent. or more above their nominal rating.

It would therefore seem that these combustion volumes ought to be more than ample and that no gain in economy should be expected from an increase in the ratio.

When boilers are suspended in battery it is best to place the supporting columns entirely outside of the setting walls, using only four columns with beams of sufficient strength to support the boilers in a single span. With standard I-beams it is possible to support in this manner three boilers of any diameter not exceeding 78 inches or two boilers of larger diameter. If the installation involves more boilers it is best to set them in separate batteries of two or three boilers each, rather than to use columns in the division walls between boilers; if it is absolutely necessary to use such intermediate columns, an air space should be left all around each one with a suitable ventilating duct to admit air at the bottom. We know of several cases where columns have been burned off or otherwise damaged when built solidly into setting walls.

Our setting plans show the proper sizes of I-beams to use for suspending boilers, together with alternate designs for both round and square cast-iron columns, structural steel H-beams and built-up columns made of plates and angles. In general it will be found that these designs are heavier than those usually employed by boiler manufacturers, but we think that these sizes are needed in order for the columns to have a strength equal to that of all other parts of the installation where it is customary to use a factor of safety of 5. Boiler columns are loaded entirely at one side and the stresses are therefore greater than when the loading is symmetrical as assumed in the tables published in structural steel handbooks. Furthermore, proper consideration should always be given to the " ratio of slenderness," a heavier section being needed for a long column than for a shorter one carrying the same load. I-beams are frequently used for columns but they are not well adapted for the purpose as the distribution of metal in the I-beam section does not make a good column design. Our Engineering Department can furnish designs for reinforced concrete columns, if desired.

Boilers having a diameter of 78 inches or less can be supported by brackets which rest upon bearing plates built into the setting walls but the suspension method is better, particularly for the larger sizes. Four brackets (two on each side) are sufficient for boiler diameters of 54 inches or less but eight brackets should be used for boilers larger than this size; the brackets should be located in pairs with a single bearing plate for each pair. Brackets at the front end should rest directly upon the plates but rollers should be used under the brackets at the rear end to permit free expansion of the boiler. As a rule, not enough care is used in setting the bearing plates with the result that a good bearing is not obtained over the entire surface of both brackets. In an extreme case the bearing may be only along one edge of one bracket. The boiler should be supported by blocking or other suitable means while the setting is being built and its weight should not be allowed to come upon the walls until the mortar has thoroughly hardened so that there will be no settling.

INSTALLATION OF BOILER PIPING, VALVES, FITTINGS, ETC.

Figure C shows a typical longitudinal section through a suspended boiler with overhanging front. Several self explanatory notes will be found on this drawing relative to the proper installation of piping, valves and other details. In addition to the items mentioned the following details should receive attention in any well planned installation.

The steam gage should be graduated at least 50 per cent. in excess of the maximum allowable working pressure and it should be piped up with a siphon, union cock, drip cock, and connection with stop valve for test gage, brass pipe and fittings being used throughout.

A water glass and three gage cocks should be used. The lowest visible part of the water glass should be at least two inches above the center of the fusible plug and the gage cocks should be located within the range of the visible length of the glass. Brass pipe and fittings, $1\frac{1}{4}$ inch size, should be used for the water connection to the water column except for small boilers where the minimum size may be 1 inch. To facilitate cleaning, plugged crosses should be used in these water connections in lieu of tees or elbows.

A blow-down pipe should be provided for the water column with a gate-valve or cock. This pipe should have a diameter of at least $\frac{3}{4}$-inch and should be connected to the ash pit or some other safe and convenient point of waste. It should be secured to the boiler front near the bottom by a pipe-clip or other suitable means.

All valves and fittings should be of extra heavy pattern if the pressure exceeds 125 pounds per square inch. In Massachusetts a State law fixes this limit at 100 pounds.

A sampling pipe for flue gas analysis and three $\frac{1}{4}$ inch pipes for draft gage connections should be placed in position when the setting is being built even though it is not expected to use them. The expense is insignificant and they may prove useful.

The foregoing description is intended to cover the more important

(Continued on page 54)

The Low Pressure Boiler Hazard.

A CASUAL glance through the bound volumes of THE LOCO-MOTIVE of the last dozen years reveals the fact that during that time illustrated accounts were published of not less than fourteen very serious low pressure boiler explosions, as distinguished from a large number of lesser accidents of this nature. One or more persons were killed in eight of these fourteen selected cases, and serious personal injury resulted in several of the remaining number.

There is not available a census or even a reasonably reliable estimate of the number of either power — or low pressure heating boilers in the United States, but as a broad guess we may assume that there are more low pressure heating boilers than power boilers, and when the number of power boiler accidents of a serious nature occurring annually is considered, it is really not remarkable that one should be able to pick out this number of conspicuous low pressure boiler explosions, notwithstanding the fact that the list from which they are gathered cannot even be considered in any sense complete.

It is popularly not appreciated as well as it might be that with the low pressure house heating boiler all the essential elements, that eventually may enter into a real boiler catastrophe, are present, and in this the kind of material of which the low pressure boiler may be constructed does not seem to be a significant factor.

There are, to be sure, several features connected with such boilers that could be mentioned to controvert this statement, such as for example the much higher factor of safety of the structure of a low pressure boiler as compared with the usual factor of safety of power boilers. Also the fact that the low pressure boiler does not get so much of the forcing that the modern power boiler may be subject to. The explosion of a low pressure boiler, however, can usually be traced, not to lack of strength of its structure, but to ignorance and carelessness with the fitting up and the attendance. Over-pressures will occur in them in cases in spite of the most complete advice of the boiler manufacturer as to how they must be connected up and regarding the best way to operate them. During the heating season just passed, reports and news items reached us of a number of serious low pressure boiler accidents. One case in point is the explosion of the boiler shown in the accompanying illustration, which, from the information obtainable, seems to have been due to faulty fitting up and possibly also carelessness in handling. The boiler was used with a hot water system of heating in a garage. It was not provided with a pressure — or altitude gage and the installation was also possessed

Exploded Heating Boiler.

of the all too common fault of not having a relief valve attached directly to the boiler. Our inspectors have occasion to belabor this latter detail very frequently in inspection reports, and to our sorrow we find in not a few instances that an unduly large amount of argument is required to convince owners that a relief valve attached directly to a hot water boiler is a paramount necessity for safety, irrespective of the fact that such boilers may have free communication with an open expansion tank.

In the case of the boiler shown in the illustration, there appears to have been good reason for believing that the absence of such a relief valve was responsible for the accident. When, as is supposed, during the very cold night preceding the accident, the connections to the expansion tank froze up, it became possible for an enormous pressure to build up in the boiler when it was fired up in the morning. One man, who stood directly in front of it at the time, was killed, and another man was seriously injured.

It may not be out of place to remind owners of heating boilers that a periodical inspection of such apparatus by experts, whose whole time is devoted to such work, is of immense importance.

Summary of Inspectors' Work for 1919.

Number of visits of inspection made	203,671
Total number of boilers examined	371,285
Number inspected internally	160,847
Number inspected by hydrostatic pressure	9,043
Number of boilers found to be uninsurable	1,042
Number of shop boilers inspected	10,548
Number of fly wheels inspected	26,980
Number of premises where pipe lines were inspected	8,046

SUMMARY OF DEFECTS DISCOVERED.

Nature of Defects.	Whole Number.	Danger- ous.
Cases of sediment or loose scale	31,599	1,783
Cases of adhering scale	47,284	1,907
Cases of grooving	2,393	271
Cases of internal corrosion	21,201	817
Cases of external corrosion	11,260	955
Cases of defective bracing	1,173	230
Cases of defective staybolting	2,428	493
Settings defective	9,423	836
Fractured plates and heads	3,473	487
Burned plates	4,836	518
Laminated plates	338	27
Cases of defective riveting	1,443	324
Cases of leakage around tubes	13,318	1,226
Cases of defective tubes or flues	19,700	6,353
Cases of leakage at seams	5,738	413
Water gauges defective	4,158	789
Blow-offs defective	5,325	1,488
Cases of low water	421	149
Safety-valves overloaded	1,134	289
Safety-valves defective	2,077	374
Pressure gauges defective	7,332	628
Boilers without pressure gauges	634	82
Miscellaneous defects	5,588	664
Total	202,276	20,603

GRAND TOTAL OF THE INSPECTORS' WORK FROM THE TIME THE COMPANY BEGAN BUSINESS, TO JANUARY 1, 1920.

Visits of inspection made	4,733,353
Whole number of inspections (both internal and external)	9,389,203
Complete internal inspections	3,695,635
Boilers tested by hydrostatic pressure	367,113
Total number of boilers condemned	27,839
Total number of defects discovered	5,284,821
Total number of dangerous defects discovered	580,620

The Locomotive

DEVOTED TO POWER PLANT PROTECTION

PUBLISHED QUARTERLY

HARTFORD, APRIL, 1920.

SINGLE COPIES *can be obtained free by calling at any of the company's agencies.*
Subscription price 50 cents per year when mailed from this office.
Recent bound volumes one dollar each. Earlier ones two dollars.
Reprinting matter from this paper is permitted if credited to
THE LOCOMOTIVE OF THE HARTFORD STEAM BOILER I. & I. CO.

THE truth of the old adage, that one is never too old to learn, we occasionally have brought home to us through our efforts of teaching firemen by mail the basic facts of combustion, so that they may recognize the difference between wasteful and economical furnace conditions. Among the recent graduates of our Correspondence Course on combustion and boiler handling was a man well past the days of prime youth. When he applied for enrollment he wrote: "I am 65 years young but willing to learn."

He did very well indeed. As regularly as clockwork came in his well-nigh perfect answers to the review questions and in just about four months he had mastered all the lessons. At the end of his studies we were pleased to extend to him a certificate which was marked "with honor grade."

Personal.

Mr. James G. Reid was appointed Chief Inspector of the Baltimore Department of this Company to fill the vacancy caused by the death of Mr. R. E. Munro.

Mr. Reid first became associated with The Hartford Steam Boiler Inspection and Insurance Company in 1909 as inspector at the Chicago Department, in which he served as an active and directing inspector. In 1917 he assumed charge of the inspection work of the local office at Detroit, and with such success that his promotion to the position of Chief Inspector was a natural step. We heartily commend him to the favorable consideration of our assured.

After forty years and one month of continuous service, Mr. A. W. Getchell, an inspector at the Cleveland Office retired on January 31st, 1920 from active duties and has taken up his residence at Santa Anna, Cal. Mr. Getchell has not severed his connection with the Company though he carries no responsibilities.

He came to Cleveland with his parents in 1853 at the age of three years. From the very beginning of his business career he was linked up with steam boilers and machinery and at the age of 26 he was Chief Engineer on the passenger steamer "Concord" plying between Chicago and Ogdensburg. In 1880 he became connected with The Hartford Steam Boiler Inspection and Insurance Company and has served it faithfully ever since. On the day of his leaving the surroundings of so many years of useful activity, one of those delightful informal meetings, at which good will and esteem come plainly to the fore, was held at the Cleveland office and Mr. Getchell was presented by his office associates with a gold chain, diamond set charm, gold knife and stickpin. We wish him many happy years to enjoy his well earned retirement.

OBITUARY

ROBERT E. MUNRO.

Mr. R. E. Munro, Chief Inspector of the Baltimore Department, died on March 29th, 1920, after a prolonged illness. He was born June 14th, 1862, and educated in Liverpool, England, being graduated from the Liverpool Institute. After serving his apprenticeship with Rollinson's Engineering Works, Liverpool, his early career was as engineer for various steamship lines, his last engagement being with the Red Star Line, on board the "Pennland," one of the largest ocean liners of her time. In 1888 Mr. Munro settled in this country and accepted the position of Chief Engineer for a large oilcloth manufacturing establishment, at Astoria, L. I., New York, remaining there until September 1891, when he became an inspector in the Baltimore office. He soon attracted the attention of the officers of the Company and in 1893 was appointed Chief Inspector of that Department, which position he held at the time of his death.

During the many years of Mr. Munro's connection with the Hartford Steam Boiler Inspection and Insurance Company, he was very highly regarded by its clients in his department and his aid and advice on engineering matters were frequently sought.

In addition to his professional attainments Mr. Munro was the happy possessor of a genial and sympathetic disposition. He made many friends and was much beloved by his associates in his office and a large circle of personal friends. Besides his widow, Mr Munro leaves three sons, two daughters and five grandchildren. To the members of his family we express our condolence in their great loss.

Mr. John McGinley, an inspector of this Company in the Phila-delphia Department since 1903, died on February 19th, 1920, at his home at Chester, Pa., after a long illness. He had been practically an invalid for about two years before his death. Mr. McGinley was born in Ireland in 1866 and for a long time followed the trade of boilermaking. His occupation at the time of entering the Company's employ was that of foreman boilermaker. He leaves a widow and three small children to whom we extend our deep sympathy in their bereavement.

The title page and index for Vol. XXXII of THE LOCOMOTIVE is now available for distribution to those of our readers who wish to bind their copies of the years 1918 and 1919. Upon application to the Home Office of the Company these title pages and indices will be furnished.

Horizontal Tubular Boiler Settings.

(Continued from page 48)

features connected with the construction of brick settings for horizontal tubular boilers and the installation of such boilers in accordance with good practice but without any unnecessary frills. As stated before, many of our new setting plans are now completed and available for distribution to our friends upon request. Any inquiries regarding special features in connection with this general subject will receive the best attention of our Engineering Department at any time.

Boiler Explosions.

(199) — A drum of a water tube boiler ruptured due to bulging out of the top of drum shell on June 1st, at the plant of the Nebraska Cement Company, Superior, Neb.

(200) — A boiler exploded on June 3rd in the Hopkins Creamery, Hopkins, Mich., killing two men and seriously injuring two others. A number of others were buried in the debris but escaped with slight injuries. The building was wrecked.

(201) — A tube exploded in a boiler on June 3rd at the plant of R. & H. Simon Silk Mill, Easton, Pa. Four men were injured, one fatally.

(202) — Two tubes burst in a water tube boiler on June 4th at the plant of the Delta Light and Traction Company, Greenville, Miss.

(203) — A crown sheet collapsed in a locomotive boiler belonging to the Kirby Lumber Co., Houston, Tex., on June 4th.

(204) — A boiler blew up on June 4th in the Belcher Saw Mill, 28 miles S. E. of Tuscaloosa, Ala., killing one man and injuring two others.

(205) — A blow-off failed on June 5th at the Home Ice Factory, San Antonio, Tex.

(206) — A boiler accident took place on June 5th on board the steamer "Kingston," Toronto, Ont., Canada. Two were injured.

(207) — A tee connection failed on a boiler mud drum on June 10th at the plant of the Davies Box and Lumber Company, Blairsden, Plumas County, Cal.

(208) — The head blew out of a drum of a water tube boiler on June 12th at Cane's pencil factory, New Market, Ont., Canada, injuring 11 persons and doing great damage to buildings and machinery.

(209) — On June 13th, while 18 miles north of Fort Worth, Tex., the boiler of a train locomotive of the Fort Worth and Denver City Railway exploded killing the engineer and fireman.

(210) — A boiler blew up on June 13th in the Deep River, Ia., electric light plant, injuring one man.

(211) — A rupture of a boiler shell took place on June 13th at the plant of the City Ice Company, Jeffersonville, Ind.

(212) — On June 13th a tube pulled out of the tube sheet in a water tube boiler of the North Star Egg Case Company, Quincy, Ill.

(213) — A tank containing carbonic acid gas exploded on June 14th in the drug store of Dr. Jessup, Diagonal, Ia., killing one and severely injuring two other persons. The interior of the store was wrecked.

(214) — The mud drum of a water tube boiler exploded on June 15th at the plant of the Pittsburg Plate Glass Co., at Kokomo, Ind.

(215) — The shell of a boiler ruptured on June 16th at the plant of Max Hahn Packing Company, Dallas, Tex.

(216) — On June 17th the shell of a boiler ruptured at the plant of the Port Blakely Mill Co., Pork Blakely, Wash.

(217) — A boiler used in drilling a well at Burkburnet, Tex., exploded on June 17th, killing one and fatally injuring two other persons.

(218) — A boiler ruptured on June 18th at the plant of the Texas Pressed Brick Company, Ferris, Tex.

(219) — A boiler accident occurred on June 19th on board the whale back steamer "Atikokan" while at the wharf, Montreal, P. Q., Canada. Three men were scalded to death.

(220) — On June 19th a boiler exploded at the saw mill belonging to John Reeves, one mile east of Brooklyn, Ill. Two men were injured, one of whom died the following day.

(221) — A boiler exploded on June 20th at the plant of the Hudson Valley Ice Co., Albany, N. Y. It shot up through the engine room roof and when returning demolished the building.

(222) — A small cast iron boiler used for heating water exploded on June 20th, in the basement of the Elk Hotel, Denver, Colo. Two persons were seriously injured.

(223) — A crown sheet collapsed in a logging-locomotive boiler belonging to Larkin Green Logging Co., Blind Slough, Ore., on June 21st.

(224) — A boiler exploded on June 21st on board the naval tender " Melville" while in tow of the Collier Orion off Colon. Six men were killed.

(225) — A 5" stop valve exploded on June 21st at the plant of the Lyon Lumber Co., Garyville, La.

(226) — A large air tank exploded on June 23rd at the Redmon Motor Car Co. garage, Paris, Ky.

(227) — Five tubes pulled out of the tube sheet of a water tube boiler on June 24th at the plant of the By-Products Coke Corporation, South Chicago, Ill.

(228) — Two sections cracked in a C. I. heating boiler on June 24th at Clark University, Worcester, Mass.

(229) — The boiler of a locomotive pulling a troop train near Omaha, Neb., exploded on June 25th. One man was seriously injured.

(230) — On June 26th a tube in a water tube boiler ruptured at the plant of Stevens and Thompson and Walloomsac Paper Company, Walloomsac, N. Y. Two men received burns, one of them died shortly after.

(231) — Two corrugated furnace flues collapsed on June 27th in a marine type boiler at the boiler house of the West Philadelphia Stock Yards, due to low water which was caused by the choking up of lower water column connection.

(232) — On June 30th a blow-off pipe failed on a boiler belonging to Woodward Creamery Company, Woodward, Okla.

(233) — Waterhammer action in a steam pipe caused the fracture of a 4" tee on June 30th at the plant of McCleary, Wallin and Crouse, Amsterdam, N. Y. This was caused by feed water entering the steam pipe through a leaky diaphragm of a feed water regulating device.

(234) — A boiler exploded on June 30th at the oil fields of the P. Welch Oil Company, Maricopa, Cal.

JULY, 1919.

(235) — The boiler of a passenger train locomotive exploded on the N. Y. Central railroad at Dunkirk, N. Y., when a rear-end collision took place on July 1st. The engineer and fireman were fatally scalded.

(236) — A tube ruptured in a water tube boiler on July 4th at the plant of Armour and Company, Chicago, Ill. Two men were seriously scalded.

(237) — A boiler exploded on July 5th at the saw mill of G. W. Henry three miles S. E. of Whitwell, Tenn. One man was killed and another seriously injured.

(238) — A boiler exploded on board the yacht "Flyer" on July 10th at Southampton, N. Y. Two men were killed and one seriously injured. The deck and upper portion of the ship were wrecked.

(239) — A steam heater exploded on July 11th at Treber's warehouse, Deadwood, S. D., resulting in serious injuries to two boys and considerable property damage.

(240) — A mud drum pulled off the nipples by which it was attached to the boiler at the plant of the Pittsburgh Plate Glass Company, Pittsburgh, Pa., on July 14th.

(241) — On July 14th, the top of a drum of a water tube boiler bulged and ruptured at the Farrell Works of the Carnegie Steel Company, Sharon, Pa.

(242) — The boiler of a locomotive pulling a heavy West Shore freight train blew up while traveling at 30 miles an hour near Kingston, N. Y., on July 15th, killing three men.

(243) — A tube burst in a water boiler on July 15th at the plant of the Lorain Steel Company, Johnstown, Pa., fatally scalding one man.

(244) — A crown sheet came down on July 16th in a track locomotive belonging to the Birdsboro Stone Co.

(245) — A header in a water tube boiler cracked on July 17th at the plant of the Industrial Works, Bay City, Mich.

(246) — On July 19th, the boiler of a threshing machine blew up at Maysville, Ky. Two men were seriously scalded.

(247) — A rupture of a boiler shell took place on July 19th at the plant of the General Ice Co., Amityville, L. I., N. Y.

(248) — On July 21st, the boiler of the locomotive pulling Union Pacific train No. 5008 blew up near Castle Rock, Utah, instantly killing three men.

(249) — A boiler exploded on July 22nd at the Banner Laundry, St. Paul, Minn., killing one and injuring ten others.

(250) — A header in a water tube boiler fractured on July 22nd at the plant of the Westinghouse Airbrake Company, Wilmerding, Penna.

(251) — On July 25th, the boiler of a threshing outfit exploded on the E. and C. Bell farm, 15 miles S. E. of Charleston, Ill., seriously injuring two men.

(252) — A boiler exploded on July 26th at the cheese factory of Thos. Anglin, Kingston, Ont., Canada, killing one and seriously injuring four.

(253) — A rupture of a boiler shell took place at the plant of the Ashland Ice and Cold Storage Company, Ashland, Neb., on July 27th.

(254) — A boiler explosion took place on July 28th at the plant of the Model Laundry, East St. Louis, Mo.

(255) — A boiler blew up on July 29th at the plant of the Heldenfels Shipbuilding Co., and landed about a quarter of a mile away. Four men were instantly killed and the property damage was estimated at over $10,000.

(256) — A gas tank exploded as it was being removed from a wagon on July 30th in front of 92 Boerum Street, Brooklyn, N. Y. One man was instantly killled.

(257) — A tube ruptured in a water tube boiler on July 30th at the plant of the Michigan Alkali Company, Wyandotte, Mich.

AUGUST, 1919.

(258) — A tube ruptured in a water tube boiler on August 1st at the plant of the Public Service Corp. of N. J., Perth Amboy, N. J.

(259) — A blow-off pipe failed on August 1st at the plant of the Wm. S. Merrill Company, Cincinnati, Ohio. One man was injured.

(260) — A tube ruptured in a water tube boiler on August 1st at the plant of Mitchell Brothers Company, Cadillac, Mich. Two men were fatally burned and one other seriously burned.

(261) — A tube ruptured in a water tube boiler on August 2d at the power house of B. F. Goodrich Company, Akron, Ohio. Four men were slightly burned.

(262) — A header of a water tube boiler cracked on August 2d at the power house of the Pascagoula Ry. & Power Company, Pascagoula, Miss.

(263) — A tube in a water tube boiler ruptured on August 3d, fatally scalding one and seriously scalding another man at the plant of the Carthage Board & Paper Company, Carthage, Ind.

(264) — On August 4th a boiler exploded at the Standard Oil Works, Richmond, Cal. Three men were seriously scalded.

(265) — A tube ruptured in a water tube boiler at the Tuberculosis Sanitarium near Cresson, Penna. on August 4th.

(266) — On August 4th a boiler belonging to a threshing outfit exploded on the farm of Mr. Owen Meyers, Paris, Ill., killing one and injuring three others severely.

(267) — On August 6th an ammonia tank exploded at the plant of the Houston Ice Cream Company, Houston, Texas. Eight men were injured.

(268) — Waterhammer in a main steam line caused the disruption of a blank flange on August 7 at the plant of the Camden Forge Company, Camden, N. J.

(269) — On August 8th a boiler exploded on the dredgeboat "John Callup" of the Missouri Portland Cement Company near Nonconnah Creek. One man was killed and four others injured.

(270) — Four sections in a cast iron boiler failed on August 9th at the Apartment House, 124 West 72d Street, New York City, N. Y.

(271) — A tube in a water tube boiler burst on August 10th at the plant of the Prudential Oil Company, Baltimore, Md.

(272) — On August 11th a tube burst in a water tube boiler on board the wooden steamer "Fort Wright," while off the coast of lower California.

(273) — An explosion occurred on August 11th with a boiler used for threshing near Weatherford, Okla., painfully injuring one man.

(274) — A boiler exploded on August 11th at the camp of the Sutter Basin Company near Knights landing, Cal. One man was severely injured and the resulting fire did many thousands of dollars' worth of damage.

(275) — On August 13th the boiler of a threshing machine at the farm of John P. Wallace, Agency, Mo., exploded. One man was fatally injured.

(276) — On August 14th a boiler exploded in the lumber mill of Coulbourne Brothers near Eure Station, N. C., killing four and seriously injuring six others. The mill was totally demolished.

(277) — On August 10th a boiler used on a tire vulcanizing machine exploded at the plant of the Bellevue Tire Company, Fremont, Ohio.

(278) — On August 17th a boiler which furnished steam for a mine pump exploded on the farm of Norman Mayberry, five miles N. E. of Greenfield, Ill., killing three children and injuring fourteen.

(279) — The head of a kier blew off on August 21st at the Chester Lace Mills, Chester, Penna., causing a damage of $5,000.

(280) — A tube ruptured in a water tube boiler on August 24th at the plant of the Dubuque Electric Company, Dubuque, Iowa.

(281) — On August 25th a boiler exploded in the Toledo Canning Company's factory, Toledo, Ohio. Two men were severely scalded.

(282) — A blow-off pipe failed on August 26th at the plant of the Moody Bible Institute, Chicago, Ill.

(283) — On August 26th a boiler exploded in the sawmill of LeRoy Marbelle near Marlo, Wash. Two men were killed and two others were injured.

(284) — On August 27th a boiler used with a gas rigging outfit exploded on the farm of Fred McMannus three miles east of Carthage, Indiana. Two men were injured.

(285) — On August 28th a battery of three boilers exploded at the sawmill of W. J. Swann, at Stonewall, N. C. Three men were killed and several others more or less severely injured. The mill was demolished.

(286) — On August 29th a boiler belonging to the Shelby Oil Company near Fairmont, W. Va., exploded killing one man.

(287) — On August 29th the plant of the Green River Electric Light, Water & Ice Company, at Calhoun, Ky., was destroyed by a boiler explosion. The only man that was in the plant at the time was severely injured.

(288) — A tube in a water tube boiler burst on August 30th at the plant of the Aetna Portland Cement Corp'n, Fenton, Mich. One man was severely scalded.

SEPTEMBER, 1919.

(289) — A tube of a superheater failed on September 1st at the plant of the Texas Power & Light Company, McKinney, Tex.

(290) — A tube in a water heater boiler ruptured on September 1st at the plant of the Hooker Electric Chemical Company, Niagara Falls, N. Y.

(291) — Three sections cracked in a cast iron boiler on September 1st at the apartment house, 50 West 67th St., New York City, N. Y.

(292) — On September 2d a hot water tank exploded at the home of Paul H Cromelin,, Hackensack, N. J.

(293) — Five headers cracked in a water tube boiler on September 2d, at the plant of the Consolidated Safety Pin Company, Bloomfield, N. J.

(294) — A tube sheet collapsed in a boiler on September 2d, at the plant of the Beacon Tire Company, Beacon, Dutchess County, N. Y.

(295) — An elbow on a blowoff pipe failed on September 3d at the plant of the Bailey Wall Paper Company, Cleveland, Ohio.

(296) — On September 4th a boiler exploded at the plant of the Polar Wave Ice Plant, St. Louis, Mo. Two men were severely injured, and the property damage was estimated at $5,000.

(297) — A tube of a water grate under a boiler failed on September 6th

at the plant of the Central Cotton Oil Company, Jackson, Miss. One man was
seriously scalded.

(298) — A logging locomotive boiler of the Kirby Lumber Company
exploded on September 6th, at Silsbee, Tex., killing one man and injuring two
others.

(299) — On September 9th a rupture of a boiler shell took place at the plant
of the Rockwood Brewing Company, Rockwood, Pa.

(300) — On September 10th a boiler exploded at the sawmill plant of John
Fleming, Fairhope, Ala., killing three men.

(301) — A tube pulled out of a tube sheet of a water tube boiler on
September 10th at the plant of the American Strawboard Company.

(302) — A 10" steam pipe failed under 150 lbs. pressure on September
10 at the plant of the Winchester Repeating Arms Company, New Haven,
Conn. The pipe opened up over a length of about 10 feet and the resulting
damage was over 3,000 dollars.

(303) — A blowoff pipe failed on September 10th at the plant of the
Autauga Oil & Fertilizer Co., Prattville, Ala.

(304) — On September 12th a boiler of a threshing outfit blew up at
Holder's farm near Neosho, Mo., killing one and injuring one other man so
seriously that he died the next day.

(305) — A boiler exploded on September 12th at the Shaw-Batcher ship-
yard, South San Francisco, Cal. One man was killed.

(306) — A tube of a water tube boiler burst on September 13th at the plant
of the Martinsville Gas and Electric Company, Martinsville, Ind. One man
was injured.

(307) — On September 13th a tube in a water tube boiler burst at the
Corpus Christi Ice and Electric plant, Corpus Christi, Texas. Two men were
scalded, one of whom fatally.

(308) — On September 15th a boiler exploded at the Curtis mine, twelve
miles from Steamboat Springs, Colo., fatally injuring one and slightly injuring
two other men. The damage was estimated at 4,000 dollars.

(309) — The crown sheets came down, on September 15th, in three locomotive type boilers used for well drilling at the oil lease of the Cohan Estate
near Whittier, Cal.

(310) — On September 17th the boiler of a freight locomotive exploded
on the L. and N. Railroad, at Hygeia, Tenn., killing one instantly and fatally
scalding one other man.

(311) — A heating boiler exploded on September 16th in the basement of
the home of R. G. Soule, Syracuse, N. Y. The damage was estimated at 2,000
dollars.

(312) — Six headers in a water tube boiler failed on September 17th at
the plant of the Pittsburgh Plate Glass Company, Ford City, Pa.

(313) — On September 17th a boiler blew up at an oil well at Murphysboro,
Ill.

(314) — A steam line failed on September 17th at the plant of the Wand
H. Walker Company, Pittsburgh, Pa., scalding two men.

(315) — The boiler of a locomotive exploded after falling from a 45 feet
high trestle, on September 18th, at St. Louis, Mo. Two men were killed.

(316) — A tube in a water tube boiler ruptured on September 18th at the
plant of the Winona Copper Co., Winona, Mich. One man was scalded.

(317) — A number of tubes pulled loose from the tube sheet of a water tube boiler on September 19th at the plant of the National Forge and Tool Company, Irvine, Pa.

(318.) — Two sections failed in a C. I. sectional boiler on September 22d at the apartment building of the Layman Land Company, Des Moines, Iowa.

(319) — A cap blew off from a sectional (power) boiler on September 22d at the plant of the Bradlee and Company, Philadelphia, Pa. One man was injured.

(320) — A tube pulled out of a drum of water tube boiler on September 22d at the plant of the Arizona Copper Company, Clifton, Ariz.

(321) — The shell of a boiler ruptured on September 23d at the plant of the Nazareth Brick Company, Inc., Nazareth, Pa.

(322) — On September 24th a boiler exploded at an oil well near Irvine's Mills, Bradford, Pa., fatally scalding one man.

(323) — A heating boiler exploded on September 24th at the Jewish Synagogue, Cheyenne, Wyo., fatally scalding the janitor and wrecking the building.

(324) — A header failed in a water tube boiler on September 24th at the plant of W. M. Bransford, Salt Lake City, Utah.

(325) — On September 25th a boiler exploded in a sawmill belonging to Stanford Lecates at Ross Point near Laurel, Del., resulting in the death of 4 persons and serious injury to 4 others. The mill was completely demolished.

(326) — A section in a cast iron boiler ruptured on September 25th at the building of the Wildwood Apartment Company, Jackson, Mich.

(327) — The boiler of a locomotive belonging to the Northwestern Railway Co. blew up on September 26th at the roundhouse at Norfolk, Neb., killing one man and slightly injuring three.

(328) — An acetylene tank exploded on September 26th at the Vickers Shipbuilding plant, Montreal, Que., killing one man and injuring sixteen.

(329) — A blowoff pipe failed on September 26th at the plant of the J. H. Smith Grape Juice Company, Lawton, Mich.

(330) — A boiler exploded on September 26th, at the National Tank Manufacturing Company, Los Angeles, Cal. One man was seriously injured by a flying fragment of the boiler.

(331) One section cracked in a cast iron sectional boiler on September 26th at the Windsor Avenue Congregational Church, Hartford, Conn.

(332) — A tube in a water tube boiler burst on September 26th at the plant of The Central Illinois Public Service Company, Mounds, Ill.

(333) — A blowoff pipe failed on September 26th at the plant of Spies Milling Company, Preston, Mich.

(334) — A furnace collapsed in a boiler on September 27th on board the dredge "Omega" at Calexico, Cal.

(335) — A tube in a water tube boiler burst on September 27th at the plant of the Hammermill Paper Company, Erie, Pa.

(336) — A hot water tank burst in the kitchen of the home of M. A. Frazar, Brookline, Mass., on September 28th, tearing a large hole in the side of the house. The damage was estimated at 2,000 dollars.

The Hartford Steam Boiler Inspection and Insurance Company.

ABSTRACT OF STATEMENT, JANUARY 1, 1920.

Capital Stock, . . . $2,000,000.00.

ASSETS.

Cash in offices and banks	$390,221.07
Real Estate	90,000.00
Mortgage and collateral loans	1,426,250.00
Bonds and stocks	5,702,983.62
Premiums in course of collection	597,171.35
Interest accrued	107,590.44
Total assets	**$8,314,216.48**

LIABILITIES.

Reserve for unearned premiums		$3,715,903.48
Reserve for losses		175,539.16
Reserve for taxes and other contingencies		401,420.50
Capital stock	$2,000,000.00	
Surplus over all liabilities	2,021,353.34	
Surplus to Policy-holders		**$4,021,353.34**
Total liabilities		$8,314,216.48

CHARLES S. BLAKE, President.

FRANCIS B. ALLEN, Vice-President,   W. R. C. CORSON, Secretary.

L. F. MIDDLEBROOK, Assistant Secretary.

E. S. BERRY, Assistant Secretary.

S. F. JETER, Chief Engineer.

H. E. DART, Supt. Engineering Dept.

F. M. FITCH, Auditor.

J. J. GRAHAM, Supt. of Agencies.

BOARD OF DIRECTORS

ATWOOD COLLINS, President, Security Trust Co., Hartford, Conn.

LUCIUS F. ROBINSON, Attorney, Hartford, Conn.

JOHN O. ENDERS, President, United States Bank, Hartford, Conn.

MORGAN B. BRAINARD, Vice-Pres. and Treasurer, Ætna Life Insurance Co., Hartford, Conn.

FRANCIS B. ALLEN, Vice-Pres., The Hartford Steam Boiler Inspection and Insurance Company.

CHARLES P. COOLEY, Hartford, Conn.

FRANCIS T. MAXWELL, President, The Hockanum Mills Company, Rockville, Conn.

HORACE B. CHENEY, Cheney Brothers Silk Manufacturers, South Manchester, Conn.

D. NEWTON BARNEY, Treasurer, The Hartford Electric Light Co., Hartford, Conn.

DR. GEORGE C. F. WILLIAMS, President and Treasurer, The Capewell Horse Nail Co., Hartford, Conn.

JOSEPH R. ENSIGN, President, The Ensign-Bickford Co., Simsbury, Conn.

EDWARD MILLIGAN, President, The Phœnix Insurance Co., Hartford, Conn.

EDWARD B. HATCH, President, The Johns-Pratt Co., Hartford, Conn.

MORGAN G. BULKELEY, JR., Ass't Treas., Ætna Life Ins. Co., Hartford, Conn.

CHARLES S. BLAKE, President, The Hartford Steam Boiler Inspection and Insurance Co.

Incorporated 1866.

Charter Perpetual

INSURES AGAINST LOSS FROM DAMAGE TO PROPERTY AND PERSONS, DUE TO BOILER OR FLYWHEEL EXPLOSIONS AND ENGINE BREAKAGE

Department.	Representatives.
ATLANTA, Ga., 1103-1106 Atlanta Trust Bldg.	W. M. Francis, Manager. C. R. Summers, Chief Inspector.
BALTIMORE, Md., 13-14-15 Abell Bldg.	Lawford & McKim, General Agents. James G. Reid, Chief Inspector.
BOSTON, Mass., 4 Liberty Sq., Cor. Water St.	Ward I. Cornell, Manager. Charles D. Noyes, Chief Inspector.
BRIDGEPORT, Ct., 404-405 City Savings Bank Bldg.	W. G. Lineburgh & Son, General Agents. E. Mason Parry, Chief Inspector.
CHICAGO, Ill., 209 West Jackson B'l'v'd	J. F. Criswell, Manager. P. M. Murray, Ass't Manager. J. P. Morrison, Chief Inspector. J. T. Coleman, Ass't Chief Inspector.
CINCINNATI, Ohio, First National Bank Bldg.	W. E. Gleason, Manager. Walter Gerner, Chief Inspector.
CLEVELAND, Ohio, Leader Bldg.	H. A. Baumhart, Manager. L. T. Gregg, Chief Inspector.
DENVER, Colo., 918-920 Gas & Electric Bldg.	J. H. Chesnutt, Manager and Chief Inspector.
HARTFORD, Conn., 56 Prospect St.	F. H. Williams, Jr., General Agent. E. Mason Parry, Chief Inspector.
NEW ORLEANS, La., 308 Canal Bank Bldg.	R. T. Burwell, Mgr. and Chief Inspector. E. Unsworth, Ass't Chief Inspector.
NEW YORK, N. Y., 100 William St.	C. C. Gardiner, Manager. Joseph H. McNeill, Chief Inspector. A. E. Bonnet, Ass't Chief Inspector.
PHILADELPHIA, Pa., 142 South Fourth St.	A. S. Wickham, Manager. Wm. J. Farran, Consulting Engineer. S. B. Adams, Chief Inspector.
PITTSBURGH, Pa., 1807-8-9-10 Arrott Bldg.	Geo. S. Reynolds, Manager. J. A. Snyder, Chief Inspector.
PORTLAND, Ore., 306 Yeon Bldg.	McCargar, Bates & Lively, General Agents. C. B. Paddock, Chief Inspector.
SAN FRANCISCO, Cal., 339-341 Sansome St.	H. R. Mann & Co., General Agents. J. B. Warner, Chief Inspector.
ST. LOUIS, Mo., 319 North Fourth St.	C. D. Ashcroft, Manager. Eugene Webb, Chief Inspector.
TORONTO, Canada, Continental Life Bldg.	H. N. Roberts, President Boiler Inspection and Insurance Company of Canada.

The Locomotive

DEVOTED TO POWER PLANT PROTECTION

PUBLISHED QUARTERLY

Vol. XXXIII.   HARTFORD, CONN., JULY, 1920.   No. 3

BROKEN CROSSHEAD ON AMMONIA COMPRESSOR.

A Mysterious Engine Accident.

ON the front cover and on the following page of this issue we are presenting two views of an accident which occurred not long ago to the steam cylinder of an Ammonia Compressor. Fortunately no one was injured.

The machine in question consisted of a single horizontal steam cylinder with two vertical compressor cylinders. The engineer had started this engine up shortly before noon of the day of the accident and, in conjunction with two other ice machines, it had been running smoothly and quietly up to the very instant of the accident. At 4:50 P. M., the engineer passed through the engine room and found everything to be operating to his satisfaction. Five minutes after he left the room the oiler, who was giving his attention to one of the other machines, heard just one crash without the slightest noise or forewarning of any trouble. He immediately ran to the throttle of the machine and shut off the steam. The engineer arrived soon after and closed the throttle valves on the other two machines.

It was found that the machine had stopped on or very close to the crank end dead center. The cross-head was broken through the wrist pin hole, and apparently it was at this point that failure first took place. A piece of the cross head shoe was found lying on the bottom guide between the two broken parts of the cross-head, which fact might be taken to indicate that the crank-pin had passed dead center and was moving toward the head end when the accident occurred.

When the cross-head failed the steam pressure carried the piston, piston rod and the half of the cross head with a high velocity toward the head end of the cylinder. The piston struck the cylinder head, drove it off and the latter was found lying on the floor immediately back of and to the right of the cylinder. The follower plate was cracked circumferentially through the bolt holes for a distance of about thirty inches and the cylinder casting, as is shown in the illustration, was fractured through the steam valve chest. Upon further examination the piston was found to be drawn off the rod, leaving a space of 3/16", and on the rod a collar of metal had been rolled up as though the piston had been driven onto the rod. This condition was probably caused by the following series of events: When the cross head failed the steam pressure carried the piston up against the cylinder head which caused it to let go and the cylinder casting to fail as shown. Before the head let go, however, the inertia of the rod and the half of the cross-head attached to it had carried the rod into the piston which caused the collar of metal to be rolled up. The force of these two actions, however, was not sufficient to bring the moving mass to rest so that, after the cylinder head blew out, the parts continued their

VIEW OF HEAD END OF CYLINDER.

motion until the nut on the cross-head end of the piston rod brought up against the stuffing box on the crank end of the cylinder. While this brought the rod and cross-head to rest the remaining inertia of the piston caused it to be carried off the rod as mentioned above.

No satisfactory explanation has been given of the cause of this accident. The steam pipes were well drained and the traps found to be in good working order. The firemen reported that the water line in the boilers at the time of the accident was what they were accustomed to call a low one. There does not seem to have been any possibility therefore, for water to have gotten into the steam cylinder.

The case serves to show that any engine may be subjected at some time to forces of unknown origin which may lead to a more or less severe accident and it furthermore emphasizes the need for engine insurance.

New Method of Connecting Return Pipes to Cast Iron Heating Boilers.

TO maintain a safe water level in heating boilers in which the condensation flows back by gravity would appear to be a simple matter. The steam in such boilers is raised slightly above atmospheric pressure and rushes to the radiators where the heat is abstracted, and consequently the steam becomes water which then flows back to the boiler. No water is lost in the process except a very small percentage that may leak out of the valves and fittings but, barring that negligible amount, the water level should not change appreciably since the same amount of water enters the boiler through the return pipe as is evaporated off into the heating system through the steam pipe.

In actual practice, however, there are a number of features present in the average steam heating installation which may cause the water level to temporarily become dangerously low resulting in over-heating of the highest parts of the heating surface and, in the case of cast iron boilers, when the water subsequently rises to the proper level again and covers these over-heated parts they usually crack.

As an illustration we will take an installation of two boilers connected to one heating system and receiving their return water through a single return pipe. With such an installation the use of check-valves on the return connections to each boiler is necessary to prevent the exchange of water between the boilers. Referring to fig. 1 it will be seen that if the check valves " a " were not present, the slightest difference in pressure on the water surfaces of the boilers Nos. 1 and 2 will cause the water to flow from one boiler into the other with the possibility of low water in one of them. It may seem a little paradox-

ical to some of the readers of this article that even a slight difference in
pressure can exist in two boilers connected to a common steam main,
but experience with low pressure boilers has shown this to be a fact,
and the cause of this difference in pressure may be partly explained
as follows:—

Within a low pressure steam system the balance of temperatures
and pressures is a very delicate one due to several causes, such as
local excessive friction of the fluid passing through the pipes and local

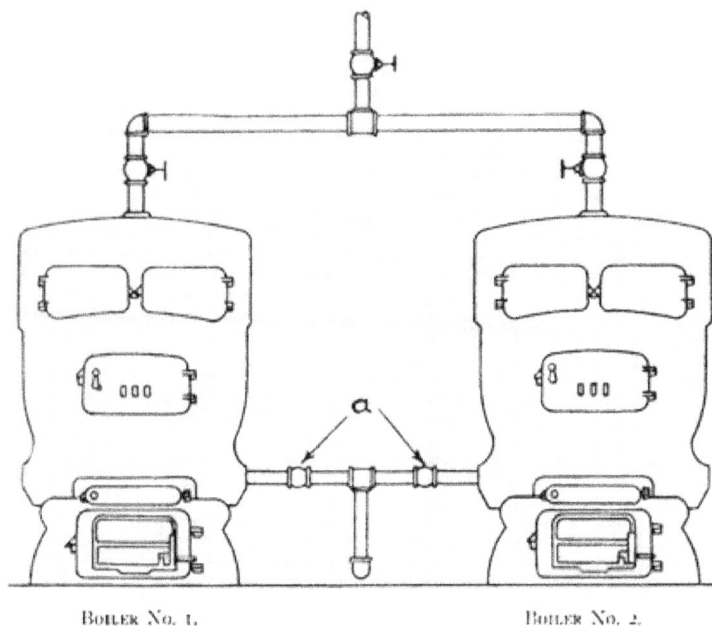

BOILER No. 1.   BOILER No. 2.

Fig. 1.

condensation, so that it is indeed hard to foretell what the actual
pressure will be at any given point. When boiler No. 1 is fired harder
than No. 2 a greater amount of steam will flow from the No. 1 boiler
than from the other. This increased flow of steam will result in increased friction in the steam pipe leading from boiler No. 1 and consequently the pressure in this boiler will rise somewhat above that
in boiler No. 2.

The pressure difference need not be very much to cause a considerable variation in water level. With the water in the boiler at a

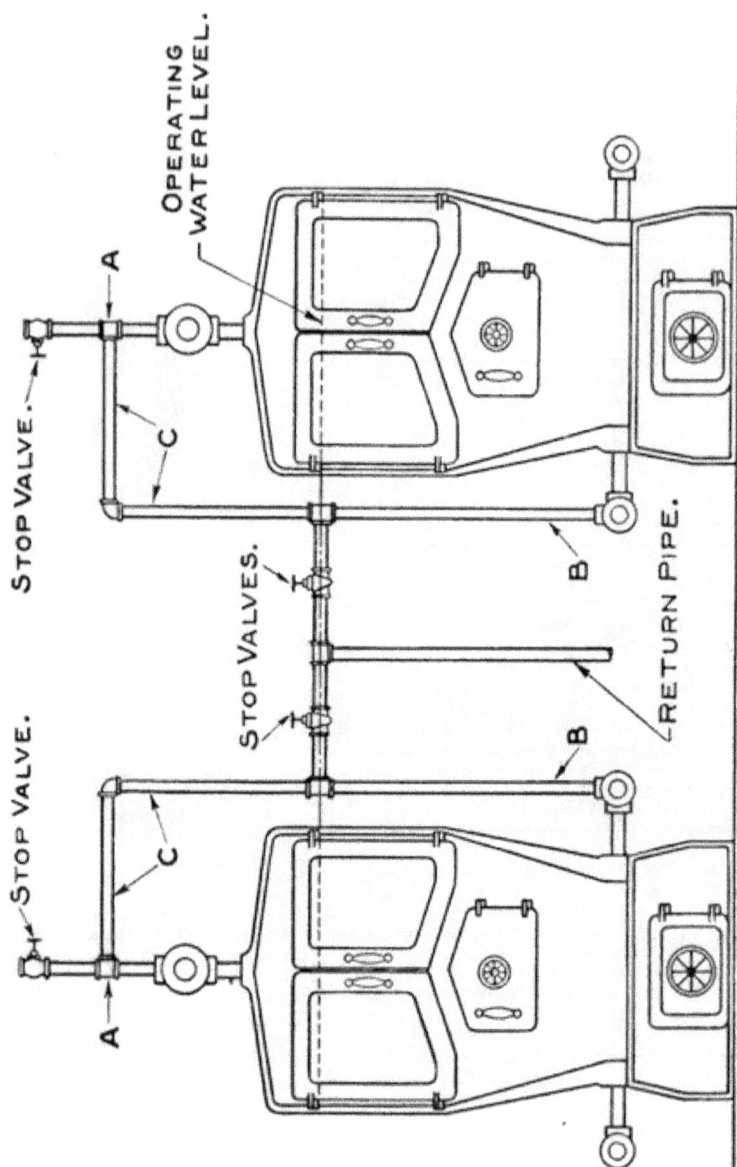

OPERATING WATER LEVEL.

A

STOP VALVE.

STOP VALVE.

C

STOP VALVES.

B

RETURN PIPE.

B

C

A

STOP VALVE.

STOP VALVE.

Fig. 2.

temperature of 220 degrees Fahrenheit a pressure difference of ¼
of a pound will cause the water level to differ approximately 7¼
inches, so that with the water only a few inches above the highest part
of the heating surface it will be seen how a slight unbalance of pressure between two boilers may cause dangerously low water in one of
them.

With the return water entering the lowest part of the boilers, as is
customary, the use of check valves is also necessary to prevent the
water from backing out of the boilers into the heating system through
the return pipe when, for instance, the stop valve on the boilers (if
present) or else the steam-valve on every radiator of the system might
be closed. The use of check-valves on the return pipes, however, is
far from being an ideal means to guard against these irregularities
for the reason that to lift a check-valve a certain amount of head
is necessary so that the operation of the check-valve itself involves a
certain amount of unbalance in the system and, since it is a practical
impossibility to get two check-valves that will act on exactly the same
pressure, the chances for a variation in water level in two boilers
getting their water through check-valves from a single return pipe are
very great. Moreover, the effect of lifting the check-valve requiring
the least pressure will tend to prevent the other check-valve from
lifting.

This problem is as old as the art of low pressure steam heating
itself and the lack of a proper solution of it thus far has undoubtedly
been the direct cause of many a cracked section in cast iron heating
boilers.

We now come to the main point of this article, the foregoing being
all by way of introduction to a new scheme of attaching the return
pipes to heating boilers as announced in the title. This new method
of piping the returns is shown in fig. 2. It will be noted that no check-
valves are used. The returning water from the heating system is
carried to a point at about the water level of the boiler although it
actually enters the boiler at the usual location at the bottom. It would,
of course, be possible to carry the water into the boiler directly at
about the water level, but this is not desirable with the average cast
iron heating boiler, for the reason that it may cause too great a temperature difference at this level in view of the fact that the returning
water can be quite cold.

However, with the arrangement as shown, it becomes necessary
to install a steam equalizing pipe (marked " c " in fig. 2.) to the highest
part of the return pipe, and this equalizing pipe must be of ample size

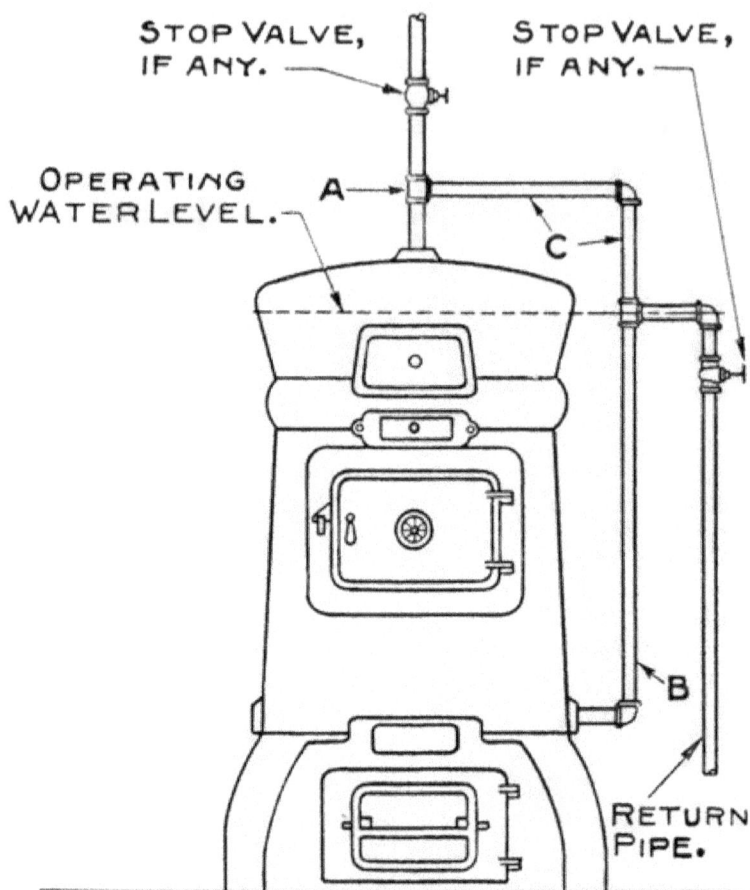

STOP VALVE,
IF ANY.

STOP VALVE,
IF ANY.

OPERATING
WATER LEVEL.

A

C

B

RETURN
PIPE.

Fig. 3.

and with as few turns as possible in order to eliminate friction, but
with such a pipe installed as shown water cannot be backed out of the
boiler below the normal water level and the baneful effect of the
installation of check-valves is thus done away with.

For the guidance of our readers, who desire to make this change
in existing installations, we would state that the minimum size of
pipe for equalizing pipes " C " should be as follows :—

Grate Area	Size of Pipe
4 square feet or less - -	1-½"
4 square feet to 15 sq. ft. - -	2-½"
15 square feet or more - -	4"

These figures apply to equalizing pipes " C " only. The return pipes and pipes " B " which are normally filled with water may be of any size which modern practice prescribes as proper for feeding boilers.

Also with a single boiler installation this method of attaching the return pipe can be used to advantage and the check-valve omitted. (see fig. 3.) It will be seen that even though there is no check-valve. water cannot be forced out of the boiler lower than the proper level, if, for instance. the stop valve on the boiler should be closed inadvertently or in case the valves on all radiators in the system should be closed. Under such conditions in a single boiler installation all the steam generated would need to escape through the equalizing pipe and the return pipe. which, of course, would produce a rattling noise, but this noise would be a real safety alarm since the operator of the boiler would be warned thereby that something is wrong and that the boiler needs attention. Under such conditions the evaporation in the boiler should be stopped by properly checking the fire so as to stop the flow of heat.

The correctness of principle of this method of piping the returns has been well demonstrated by test, so that we have no hesitation in recommending it to any owner of cast iron boilers with separate returns who desires a better safeguard against cracking of sections.

The Advantages of Co-operation Between the Boiler Manufacturer and Insurance Company.*

S. F. JETER, Chief Engineer.

CO-OPERATION as a slogan has been overworked during recent years. It has been advocated in many instances by those who apparently had no conception of its meaning, or if they did have. it was not their intention to really co-operate. The derivation of the word indicates that it means to " work with others." Many have used it apparently with the idea that it meant to " work for others." It should be understood that co-operation necessarily involves the viewing of the subject from the other fellow's standpoint, as well as your own. There are many ways in which the boiler manufacturer and the insur-

* Paper read at the annual meeting of the American Boiler Manufacturers Association at French Lick Springs, Ind., May 31, 1920.

ance company can work with each other, or co-operate, and the interest of both be served.

Insurance companies are frequently confronted with claims due to minor accidents. If in making repairs in cases where an insurance company is involved, the boiler maker would keep the cost of the repairs that were actually necessitated by the accident, separate from any general repairs that might be made at the same time, it would be very advantageous to the insurance company as well as the assured who is the boiler maker's patron. With a proper division of the items the possibility of controversy between the insurance company and its patron is avoided. Such co-operation should not cause the slightest friction between the boiler maker and his customer. All that is needed is that he so keep track of the items of repair that he will be in a position to make a correct division of the cost if called upon to do so. There have been many instances where co-operation of the boiler maker in such cases has been of the greatest aid to both the insurance company and its customer. However, there have been many such cases, more especially in connection with the small boiler maker, where lack of co-operation, if not actual antagonism, has been extremely detrimental.

The furnishing in case of explosions of prompt and proper bids or estimates by the boiler manufacturer covering replacements or repairs required is a class of co-operation that is welcomed by the insurance company.

The boiler manufacturer can of course aid the insurance company in informing his patrons or prospective patrons of the service that may be secured through the insurance company. Boiler insurance has been so widely advertised in one way or another that it would seem there could be no user of steam who was unacquainted with the facts. However, there appear to be many boiler purchasers who do not yet realize the service they can secure from boiler specialists to look after their boilers from the cradle to the grave, or rather, from the plate mill to the scrap heap, even though the termination of the existence of their boilers is not of a more serious or spectacular nature. A suggestion from the manufacturer for inspection during construction, with its accompanying policy of insurance or certificate of inspection by an organization with which the manufacturer is not allied, cannot help but impress the purchaser with the manufacturer's good faith in meeting his contract obligations. I feel sure that the advantages of the service and co-operation rendered by the insurance company in this respect are fully appreciated by the members of the American Boiler Manufacturers Association.

The insurance company can be a distinct aid to the manufacturer in seeing that the workmanship in his shop is kept up to the standard he desires to maintain. A manufacturer can, and does, of course, instruct his own employes as to the grade of work he desires to turn out and may institute a form of inspection service through his own employes for the purpose of maintaining this standard. However, an employee, particularly if he is faithful, cannot refrain from keeping in mind what he considers to be his employer's best interest. Notwithstanding the fact that some features of workmanship may not come up to the mark set, the faithful employee will always consider the cost involved before rectifying the trouble. The shop employee, as a rule, is not in a position to make a decision in such cases for the best interest of his employer. Often he will decide that the rectification of a mistake is unwarranted, owing to the cost involved, when the officials of his company would have decided that the cost should not be considered. The inspector of the insurance company, of course, is not directly concerned with the cost of rectifying a mistake. If a mistake is judged of importance enough he will condemn the work outright or, if of more minor importance, the inspector may call it to the attention of the proper shop or company official who can decide with the inspector whether a change should be made or not.

Another feature that is a detriment in securing adequate inspection by a representative who is paid by and under the direct control of the manufacturer is this: mistakes are often made, which, if brought to the attention of the shop or company officials, are liable to cause a reprimand. It is natural, under such conditions, that an employee dislikes to be the cause of the reprimand of a fellow employee. For the same reason it is desirable that an insurance inspector, while enjoying the good will and respect of shop employes, should not be on too intimate terms with them. This is a feature that the insurance company attempts to guard against in all cases.

The insurance company can co-operate with the manufacturer and his employes in determining if the standards for workmanship which have been established are in accord with those maintained by the better shops doing a similar class of work. Co-operation of this kind is very advantageous and tends to spur the employes to their best efforts to see that others do not exceed the grade of work which they turn out.

Another feature in which the insurance company has become extremely valuable to the manufacturer is in co-operating with him in meeting the legal requirements for construction where boilers must be built to come under boiler laws. This is becoming more important each day, since the states having boiler laws governing construction are constantly increasing in number. It was this tendency for boiler laws to become more common that caused the manufacturers and insurance companies to co-operate in an effort to secure uniform rules governing boiler construction. The organization of the Uniform Boiler Law Society, which is sponsored by your Association, and the introduction of the A. S. M. E. Code, the success of which is mainly due to the hearty support and unceasing labors of some of your members, are two tangible evidences of co-operation between the boiler manufacturer and the insurance company. Along this same line the insurance company can be of the greatest aid to the manufacturer in checking his designs before commencement of work to see that the requirements of the law are met. Our Company is constantly called on by the manufacturer to furnish this class of service.

Another way in which the insurance company can co-operate with the manufacturer is in securing modifications of rules or regulations that may tend to work a hardship on either party without compensating advantages to the steam user or public. I feel sure that those of your members who are directly concerned with the Code work of the A. S. M. E. can vouch for the thorough co-operation which exists between the two interests in connection with such features.

Inflammability and Explosion of Ammonia.

FROM time to time one hears of explosions of ammonia compressors where the compressor heads have been blown out, and in some cases fire has resulted. These accidents have been referred to as " mysterious," as pure anhydrous ammonia will not burn or explode.

A refrigerating plant does not have to be run many years for the system to become impregnated with gases other than pure anhydrous ammonia, especially if from lack of knowledge and experience the plant is not properly handled. Regardless of knowledge or experience it is impossible to keep the system at all times clear of gases other than pure ammonia.

A new installation of a refrigerating or ice-making system that has never been in operation for a minute is never started with strictly pure ammonia, for in the first place the ammonia itself is not always absolutely pure, and in the second place it is impossible to pump a perfect vacuum in the coils before the ammonia is charged into them. After the plant is in operation and it becomes necessary to pump out a coil or set of coils for repairs, there is always a small amount of air left behind, as there never was a compressor made perfect enough to entirely expel the air or pump a perfect vacuum.

The air left behind in this way is of no practical consequence in the operation of the plant, but if an accumulation of air and other impurities such as oil vapor, decomposed ammonia (nitrogen and hydrogen) and many other gaseous impurities is allowed, there is no knowing what composition these mixtures will form, or at what temperature they are likely to explode; and as such explosions have occurred at various places throughout the country, causing serious damage to life and property, it behooves the operating engineer to think before allowing the plant under his supervision to accumulate too great a charge of impurities or to allow the discharge line and compressor to run at too high a temperature.

New York City has passed an ordinance allowing the fire department to open up connections from a refrigerating system and blow the entire charge of ammonia into the sewer in case of fire, the valves being located in a locker outside the building, the ammonia pipe lines discharging into water and then into the sewer.

The question of inflammability and explosion of ammonia has been definitely determined by Arthur Lowenstein and his associates, R. J. Quinn and S. Drucker, who conducted research and experimental work along these lines extending over a period of several years, the chief results of which are quoted herewith from a paper written by Mr. Lowenstein and read before the Chicago branch of the American Society of the Brewing Technology, March 29, 1916:

"Briefly, however, it may be stated from these experiments that ammonia gas, mixed with air in certain proportions, when ignited, will propagate a flame. Special apparatus has been designed which enables us to determine the conditions under which ammonia will burn, when all known mixtures with air are brought together under suitable conditions; also known mixtures of oxygen and ammonia.

These tests were made in steps of 0.5 per cent., using mixtures varying from 1 to 100 per cent. of ammonia in air, both dry and saturated with water vapor. The results under these conditions were that no visible form of burning was evident until the region of 11 to 13 per cent. of ammonia was reached. It was found that a small yellow flame was produced at 11 per cent. ammonia in air, which increased in size with increase of ammonia content until at the proportion of 13.25 per cent. of NH_3 the burning was complete. At this concentration a yellow flame completely enveloped the glass containing vessel and the combustion was sufficiently violent to shatter the vessel."

Another set of experiments was made by using an electrically heated incandescent platinum wire in place of the spark. "The percentages of ammonia in air were increased up to 19.58 per cent., at which point the mixture violently exploded upon ignition. . . . . Under these conditions mixtures of between 19.58 per cent. and 26 per cent. of ammonia in air were exploded. The explosions were very violent, sufficient pressure being generated to shatter the glass container."

There is no danger of an explosion unless heat is applied at some part in the system to cause one. The place where this can occur is at the compressor. If the compressor is kept in good condition there is little danger, but if there are a set of leaky rings, which may be broken or otherwise, causing the piston to churn the gas back and forth until the temperature reaches a point where the solder will melt and run out of the discharge connections, there is a possibility of one more "mysterious explosion" and if the engineer is lucky, he will be called upon to explain why.

There are many other reasons for the compressors running hot, and the engineer will find that there will be less explaining if he will see to it that the compressor is run cool at all times; there will be a better lubricating effect, less wear to the compressor cylinder, less expense of upkeep, longer life of the machine and a longer life to the engineer.

There was one experience that the writer is always rather timid and backward about mentioning. It happened at the Cudahy Packing Company's plant at Kansas City. On passing through the engine room, he noticed a cherry-red spot on the side of a compressor piston rod about the size of a silver dollar. This spot was passing through the packing and into the cylinder, and when the machine was shut down and the rod allowed to cool, the metal was pulled away and roughed up so that the rod had to be filed off before starting the machine.

This condition was caused by the packing being allowed to remain in service until too hard, and the red spot was probably caused by a soft spot in the metal; anyway, it was rather a freak condition and one that would not have been "healthy" if an explosive mixture of impurities had been in the system.

As a final precaution against fire in case of a possible explosion, the engineer should insist on the elimination of arc lights, open gas jets or any other form of inflammable light in the room where the refrigerating or ice machines are located. *B. E. Hill in " Power."*

A Hair-Breadth Escape.

IT has been very appropriately said that the number of boiler explosions that are prevented through the advice of inspectors can never be known. This fact, while fortunate for the community at large, detracts a great deal from the credit that properly belongs to the inspector if appearances are taken as the only criterion in judging the results of his labors. Every now and then however we meet with a situation where surely an explosion would have occurred involving heavy death toll and enormous property damage were it not for the inspector's timely advice based on his intimate familiarity with steam boilers and their appendages.

A short time ago the manager of a news-printing establishment in a Middle-Western city telephoned our local representative stating that the safety valve on their boiler was blowing continually but that they were unable to raise the pressure high enough to operate the engine. He further said that their engineer had attempted to adjust the safety valve until he had screwed it down to a point where the spring broke. The boiler had then been cooled down and a new safety valve set at the desired pressure was installed but this gave no better results as regards pressure. The new valve continued to blow and still not enough pressure to run the engine.

As the printing of the first edition of the newspaper was being considerably delayed for lack of motive power, they desired to know whether it would not be all right to plug the safety-valve opening and station a man at the steam gauge to guard against over pressure until the day's work was finished. Our representative, sensing the danger and instinctively, as it were, feeling what was wrong with the installation, requested that they immediately bank the fire and await the arrival of an inspector.

Upon looking over the installation the inspector found the steam gauge on the boiler registering 15 lbs. pressure and the engineer called his particular attention to the fact that the gauge in the engine room also showed 15 lbs. Two steam gauges registering the same. Was not that proof that the trouble must be elsewhere than with the gauges?

Our inspector found however that the two steam gauges were connected to a single pipe which led to the upper water column connection and that this connection was choked which prevented the steam gauges from indicating the actual boiler pressure. It is quite probable that the amount of over-pressure, at the time that the safety valve had been screwed down until the spring broke, was very great and in fact it may have been very close to the bursting pressure of the boiler. At any rate we feel that the persons that were near the boiler at the time the engineer was experimenting with the safety valve are to be congratulated with their escape from a violent death. As the boiler was located in the basement of a four story building filled with people, the loss of life would undoubtedly have been very heavy.

It is really strange that any one, with even a slight knowledge of the make-up of a spring-safety-valve can imagine that such a valve will all of a sudden begin to relieve the boiler at a much lower pressure than that at which it is set, but somehow there appears to be a tendency in the human mind to regard the indications of the hand on the steam gauge dial as absolute. When a marked difference shows up between the pressure indicated by the steam gauge and the relieving pressure of the safety valve, the first inclination seems to have been, in a number of unfortunate accident cases, to go for the safety valve.

We cannot bear down with too much emphasis on the warning to **let the safety-valve adjustment alone** after it has once been set to relieve at the desired pressure, unless of course, a small change in the working pressure is expressly wanted under normal circumstances. One of the principal functions of the safety valve is to take care of just such conditions as the derangement of the steam gauge. Blind confidence in the steam gauge has been the direct cause of a number of boiler explosions on record and these explosions have been particularly violent. This, no doubt, is due to the fact that a very high pressure may be reached when the safety valve is overloaded and consequently an enormous amount of energy is stored up within the boiler.

One of these unfortunate cases was the explosion that took place on the U. S. gunboat "Bennington" on July 21st, 1905, while at anchor in the harbor of San Diego, Cal. In this instance a boiler was fired up from a cold condition. After the water in it had been brought to boiling temperature and the air had been blown out by the first formation of steam, one of the fire-room crew was ordered to go on top of the boiler and close the air cock. This he did, but it appears that he also closed the cock on the outlet leading to the steam gauge so that it failed to register the rapidly forming steam pressure.

(Continued on page 87)

Summary of Boiler Explosions.
1918-1919.

IN accordance with our usual custom we present herewith a table showing the total explosions, persons killed, injured, and the total of killed and injured for the past calendar year of 1919. In addi-

SUMMARY OF BOILER EXPLOSIONS FOR 1919

Month.	Number of Explosions.	Persons Killed.	Persons Injured.	Total of Killed and Injured.
January	54	15	16	31
February	44	20	39	59
March	46	9	13	22
April	34	9	13	22
May	20	8	9	17
June	36	21	28	49
July	23	19	24	43
August	31	18	59	77
September	57	26	47	73
October	46	24	30	54
November	58	13	28	41
December	77	4	36	40
Totals	526	186	342	528

SUMMARY OF BOILER EXPLOSIONS FOR 1918

Month.	Number of Explosions.	Persons Killed.	Persons Injured.	Total of Killed and Injured.
January	84	18	63	81
February	45	12	48	60
March	37	8	29	37
April	25	11	19	30
May	17	7	15	22
June	19	13	19	32
July	21	11	29	40
August	36	16	17	33
September	30	19	29	48
October	43	8	34	42
November	46	4	9	13
December	46	3	29	32
Totals	449	130	340	470

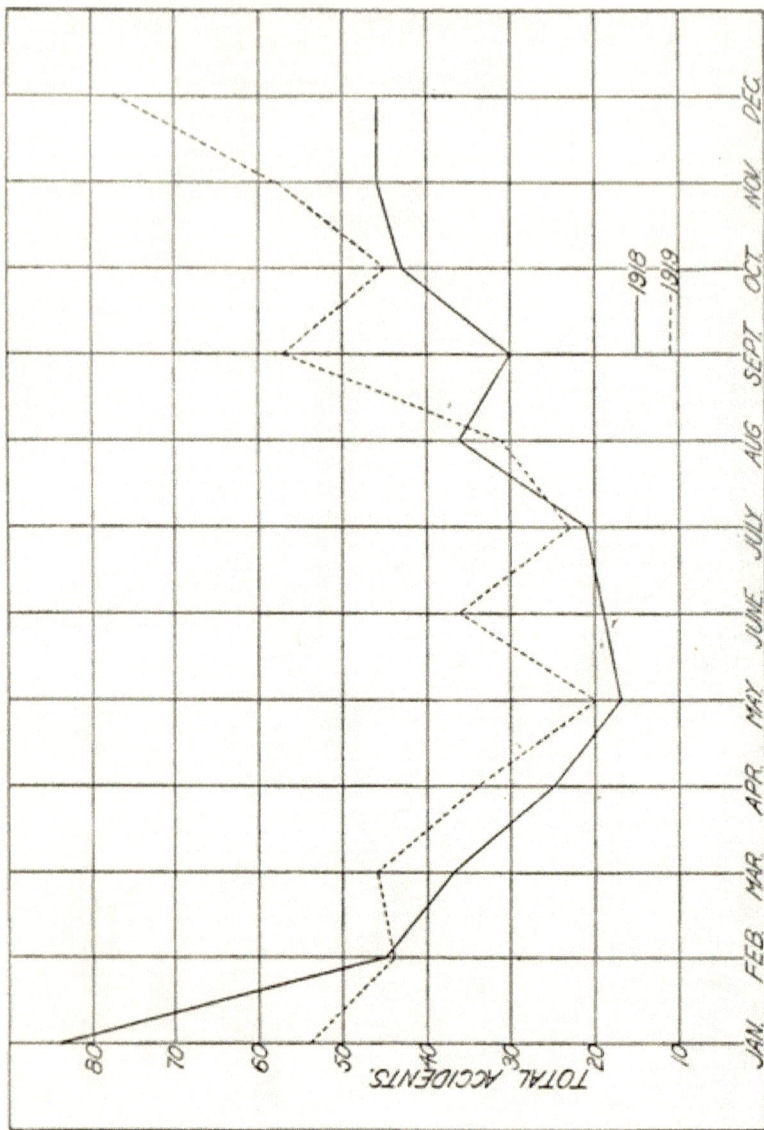

CHART OF BOILER ACCIDENTS.

tion we give the same data for the year of 1918 which was inadvertently omitted last year.

It should be understood that this company makes no claim of completeness for these lists. The information contained in them has been obtained from our own files and from such information as could be gleaned from the newspapers. There is little doubt that the papers have not reported all the accidents that have occurred and there is further certainty that a complete survey of all press reports of explosions has not been available for our use.

As a matter of additional interest we have charted the total accidents on the opposite page. The solid line on this chart represents the calendar year of 1918 and the broken line is for 1919.

Probably the most striking feature of these charts is the fact that both lines show a very marked depression during the summer months and a rise in the winter. As would very readily be supposed this decrease is effected to a considerable extent by the large number of heating boilers in use during the winter months.

We have made a further study of the data given above but for lack of space we are unable to present the results in this issue. We hope to do this in the very near future however. One of the purposes of this study has been to see what effect the heating load has upon this increase in accidents during the winter months. It is of course very difficult to say just how many of these troubles have come from the boilers which have been added to care for the heating load except in those cases where the boiler is clearly defined in that class.

Separating these clearly defined cases from the others we find that approximately one half of the rise of the chart during the Winter months is due to such cases. There are many instances, however, where additional boilers have been used in isolated plants during the winter because of the greater demand for steam and our source of information usually classifies these as being on a power load rather than a heating load. The rise in the number of accidents during this period is then in all probability due almost entirely to the heating load.

The sudden irregularities of the chart are only what we would naturally expect to find in any statistics of this nature. It is the general trend of the curve which is of the greatest interest, however, as it indicates the condition we may expect to find.

The Locomotive

DEVOTED TO POWER PLANT PROTECTION

PUBLISHED QUARTERLY

HARTFORD, JULY, 1920.

SINGLE COPIES *can be obtained free by calling at any of the company's agencies.*
Subscription price 50 cents per year when mailed from this office.
Recent bound volumes one dollar each. Earlier ones two dollars.
Reprinting matter from this paper is permitted if credited to
THE LOCOMOTIVE OF THE HARTFORD STEAM BOILER I. & I. Co.

SAFETY against accidents and personal injury in the power plant
can best be regarded as a commodity with a more or less definite
purchase value. The post-war period with its tremendous increase in the cost of all labor and material has put a severe crimp
in the purchasing power of the dollar, so that, while the industries
are still quite rushed, a good many projects of much needed expansion and improvement of the plant are being held up until the tide
of costs recedes somewhat.

The prevailing disposition, however, so far as we are able to
observe, is to place in the same category investments in additional
plant facilities and safety-improvement and we desire to caution
most urgently against this tendency. There is, of course, a distinct
difference between these two classes of investment. While the returns on plant-expansion may be quite obvious, the returns on a
safety-improvement, though very real, are rarely apparent. Those
who do invest wisely in safety-improvement can usually only realize
its true worth when accidents occur in plants of others who lacked
their foresight and prudence.

The insurance specialist is occasionally privileged to have a grand-
stand view of these matters. Not long ago a case of false economy
came to our notice which may well serve as an example that safety-
improvement is a good investment at almost any price. At a certain
plant the insurance on boilers and digesters had been carried by the

" HARTFORD " until a short time ago, when, because our recommendations for necessary improvement went unheeded, we felt obliged to cancel the insurance. Some time later one of the vessels we had criticized exploded, seriously wrecking the plant. As it happened, the owners had at the time about completed negotiations for the sale of the property for a half million dollars. The sale fell through because the owners could not deliver a going concern and so instead of a half million dollars they have on their hands today only a badly wrecked plant.

A comparatively small amount of money, invested at the right time, might have prevented this serious accident and the subsequent embarrassment of the owners.

Personal.

Mr. Frederic H. Kenyon, Special Agent at the Home Office for the past ten years, has been promoted to the position of general agent in charge of the Home Office Department, which position recently became vacant by the resignation of Mr. F. H. Williams.

Mr. Kenyon has been connected with the underwriting department of this Company for the past ten years during which time he has formed very favorable acquaintances with the Company's agents and assured. He is well acquainted with this territory so that his success as General Agent is assured.

Mr. H. J. Vander Eb has been appointed Superintendent of the Engine Department of this Company, and will devote his attention in the future to the mechanical details connected with the insurance of engines and fly-wheels.

Mr. Vander Eb came to this Company in January 1912 and has been employed in the mechanical department at the Home Office continuously since that date. His thorough engineering education and long experience as a practical engineer as well as in machine design make him well qualified for the work he is now undertaking. His previous duties with the company have been directly in line with this work and thus he is fortified with a fund of valuable experience and information.

Mr. George H. Stickney has been appointed Superintendent of the Boiler Department of this Company and will devote his attention to the mechanical details connected with the insurance of boilers.

Mr. Stickney came to the company in 1914 and served in the capacity of an inspector in the Boston Department for a period of three years. In 1917 he entered the United States Navy as Lieutenant and acted as Chief Engineer of the transport U. S. S. Pensacola. Since his release from active duty in June 1919, Mr. Stickney has been connected with the Mechanical Department of the company at the Home Office.

Announcement.

THE Hartford Steam Boiler Inspection & Insurance Company takes pleasure in announcing that it has appointed Mr. William D. Halsey editor of this publication. Mr. Halsey is a graduate in mechanical engineering of Swarthmore College, 1912. Following his course in that institution he had practical experience as maintenance and designing engineer with the Sharples Separator Company and as assistant maintenance engineer of the Baldwin Locomotive Works at their Eddystone Plant. For the past five years he has been professor of mechanical engineering at the George Washington University, from which institution he comes to our company. During the war Mr. Halsey was with the National Advisory Committee for Aeronautics, giving his attention especially to the design of Diesel engines for aircraft.

We believe Mr. Halsey is thus by his education and training particularly qualified to have charge of this periodical of ours. He brings to the work both a theoretical and practical knowledge of mechanics and an experience in teaching which fits him to give clear and interesting expression in his technical writings. Mr. Halsey's editorial duties will commence with our next number, and we are confident that under his administration THE LOCOMOTIVE will increase in interest and usefulness, maintaining the prestige it has had for over fifty years as the prominent exponent of power plant protection.

Since the resignation of its former editor, last Fall, the responsibility of issuing and editing THE LOCOMOTIVE has fallen on the shoulders of Mr. Henry J. Vander Eb, a member of our Engineering Department. Our company gratefully appreciates the cheerful manner in which Mr. Vander Eb has borne this additional burden and duty, and the care and intelligence he has displayed in the preparation of the several issues which have appeared under his charge. Others of our force have willingly aided, but his has been the responsibility for writing and editing material while continuing his other important work.

CHARLES S. BLAKE,

President.

Hartford, Conn., July 1st, 1920.

Obiturary.

Mr. Olaf Granberg, Vice-President of the Boiler Inspection and Insurance Company of Canada, died on April 17th, 1920. Mr. Granberg had been in poor health for a year or more prior to his death and the death of Mrs. Granberg last January no doubt hastened his own end.

Mr. Granberg was born in 1850 in Norway. He came to Canada in 1857 when his parents settled at Coaticooke, Province of Quebec, where he received his education. Later on he worked in the local blacksmith shop where his principal work was dressing the drills for a nearby quarry. Subsequently he entered a machine shop and served his apprenticeship as a machinist. From there he went to the plant of the Dominion Cotton Mills, Coaticooke, and rose to the position of Master Mechanic and later to Manager.

In 1890 he became affiliated with the Boiler Inspection and Insurance Company as an Inspector. In 1900 he became Chief Inspector for the Province of Quebec, and in 1905 he was appointed Manager of that district.

His earnestness and loyalty won for him the position of Vice-President of the Canadian Company to which office he was elected in 1917 and which office he held until he died. His death brings a deep sense of loss to all who knew him.

A Hair-Breadth Escape.

(Continued from page 80)

No one suspected that anything was wrong with the gauge and the boiler was fired heavily in order to hasten the steam pressure which according to the gauge remained at zero. To complete the chain of unfortunate circumstances the safety valves were stuck on their seats with the result that a dangerous pressure was permitted to build up and finally the boiler gave way in its weakest part.

The total number of persons killed was 62 and besides this there were 40 more or less severely injured. This unusually large number of casualties for a boiler explosion was, of course, due to the cramped conditions usually prevailing in a ship's power plant and the fact that on naval vessels men are likely to congregate in great numbers at any given point, but no doubt the very high pressure at the time was a contributing factor to this large loss of life.

Boiler Explosions.

(337.) — Two tubes pulled out and several other tubes were damaged at the Power House of the Arizona Copper Co., Clifton, Arizona, on September 22nd.

(338.) — On September 28th, a boiler exploded at the plant of the Crawford Mills, Pearl River, La. One man who was working in the mill 200 feet away was killed by a flying fragment. Six others were injured. The damage was estimated at $6,000.

(339.) — On September 29th, a boiler explosion on board the U. S. destroyer Greene, off Key West, killed one man and severely scalded two.

(340.) — On September 30th, a boiler exploded at the cotton gin of Shackleford Brothers, Attica, Ga., instantly killing one man and fatally injuring another.

(341.) — A three-inch stop valve in a steam line burst on September 30th, at the plant of the Bath Iron Works, Ltd., Bath, Me.

(342.) — A section cracked in a cast iron boiler on September 30th, at the Alcott Schoolhouse, Hastings, Neb.

(343.) — Two sections cracked in a cast iron boiler on September 30th, at the Woodfords Congregational Church, Portland, Me.

(344.) — On September 29th, a tank exploded at the rendering plant of Bell Bros., near Oakland, Ill. One man was killed.

(345.) — A blowoff tank exploded on September 30th, at the Cupples Station Light, Heat & Power Co., at St. Louis, Mo. Two men were injured.

OCTOBER, 1919.

(346.) — An inner flue on a drier collapsed, October 2nd, at the South Omaha, Neb., plant of Swift and Company.

(347.) — A door frame section cracked October 2nd, on a heating boiler in the plumbing shop of J. F. Morgan, Lynn, Mass.

(348.) — On October 2nd, a hot water boiler exploded in the Westminster Hotel, Los Angeles, Calif. The engine room and hotel lobby were very badly damaged.

(349.) — On October 2nd, a steam boiler exploded on the farm of James Stewart, near Howard, N. Y. Edward Stewart and his son were instantly killed and several men were injured, two seriously.

(350.) — On October 4th, a boiler exploded at the Johnson Ginnery, Baxley, Ga. Three men were killed and five badly injured.

(351.) — A crown sheet collapsed on a locomotive belonging to the California Barrell Company, Portland, Oregon, October 6th.

(352.) — On October 6th, an air tank exploded at the Midvale Steel Works, Philadelphia, Pa. One man was killed.

(353.) — A cast iron fitting ruptured at the Laurel Lake Mills, Fall River, Mass., October 6th.

(354.) — Eight sections cracked October 7th, in a boiler in the Baxter Hotel, Ida Grove, Iowa.

(355.) — An accident occurred October 7th, to a water column pipe on a boiler belonging to The Normanna Gin Company, Normanna, Texas.

(356.) — A blowoff pipe ruptured October 8th, at the Blue-Island, Ill., plant of Libby, McNeill & Libby. Two men were injured.

(357.) — On October 8th, an ammonia drum exploded at the plant of The Commonwealth Public Service Corporation, Mena, Ark. One man was killed and another seriously injured.

(358.) — On October 8th, two men were injured, one seriously, by an accident to a boiler in the Hotel Royal, New Haven, Conn.

(359.) — A section of a boiler cracked October 8th, at a public school in Rochester, Beaver County, Penn.

(360.) — A rupture occurred October 9th to a boiler belonging to the City of Wellsburg, West Va.

(361.) — On October 9th, an acetylene tank exploded at the repair shop of F. Franzwa, Carroll, Ia.

(362.) — On October 10th, a boiler on the steamer Chestnut Hill exploded at Philadelphia, Pa. Several men were killed.

(363.) — On October 10th, a boiler exploded at the sawmill of E. Johnson, St. Landry, La. Five men were killed and six injured.

(364.) — An accident occurred October 11th, to a boiler in a steam laundry of the Home Service Corporation, Los Angeles, Calif.

(365.) — On October 12th, a heating boiler exploded in the basement of the Freehafer Building, Reading, Pa.

(366.) — An accident occurred October 13th, to a boiler at the Oriental Laundry, Dallas, Texas. One man was injured.

(367.) — A manifold on a boiler was ruptured October 14th, in a building belonging to Ahlvins & Shreeve at Jackson & Collins Streets, Joliet, Ill.

(368.) — On October 15th, an explosion of a steam pipe at the Henry Clay colliery, Shamokin, Pa., was the cause of the death of one man.

(369.) — Two sections cracked October 15th, in a boiler of the Windsor Ave. Congregational Church, Hartford, Conn.

(370.) — On October 15th, a blowoff pipe was damaged at the plant of the Knoxville Furniture Company, Knoxville, Tenn. Two men were injured, one seriously.

(371.) — A tube ruptured in the main plant of the Milwaukee Coke and Gas Company, Milwaukee, Wis., on October 15th.

(372.) — On October 18th, a boiler exploded at the sawmill of E. C. Lautte & Sons, Carmel, La. Four men were killed and several injured.

(373.) — A tube was ruptured October 16th, at Plant No. 1 of the Michigan Alkali Company, Wyandotte, Mich.

(374.) — On October 19th, accidents to ammonia tanks in the Iowa Falls, Iowa, plant of Swift and Company caused the death of two men.

(375.) — Four sections of a boiler cracked October 16th, in the restaurant of J. W. Welch, Omaha, Nebr.

(376.) — Two sections of a cast iron boiler were ruptured October 20th, at the Hotel Shirley, Jackson, Mich.

(377.) — A tube ruptured October 21st, in the Morrison, Ill., plant of Libby, McNeill & Libby.

(378.) — A section of a boiler cracked October 22nd, in a theater belonging to the Armstrong Real Estate & Improvement Co., Johnsonburg, Pa.

(379.) — A section was ruptured October 23rd, in a boiler in the furniture plant of the Lauzon Furniture Company, Grand Rapids, Mich.

(380.) — Twelve headers were ruptured October 24th, in the power house of the Galveston Electric Company, Galveston, Texas.

(381.) — Tubes pulled out of a drum on a boiler at Station "B" of the Detroit City Gas Company, Detroit, Mich., on October 24th.

(382.) — An accident occurred October 24th to a boiler on a steam shovel belonging to the White-Barger-White Coal Mining Co., Grape Creek, Ill.

(383.) — An accident occurred October 27th, to a boiler in the New Marion Power Station of the Public Service Corporation of New Jersey.

(384.) — A tube ruptured at the Grasselli Powder Company, Wyside, Pa., on October 27th.

(385.) — A section cracked October 28th, on a boiler in the office building of Joseph G. Ryan, Maple and Cherry Sts., Danvers, Mass.

(386.) — On October 29th, a locomotive on the Southern Pacific Railroad, after jumping the track, exploded. Six persons were reported killed and one hundred and forty-two injured, although just how many of these casualties were due to the boiler explosion is not known.

(387.) — On October 29th, a valve ruptured at Plant No. 2 of Cosden & Co., West Tulsa, Okla. One man was killed and one injured.

(388.) — Four sections cracked October 30th, in a boiler in a garage belonging to the Stoddard Motor Car Co., Springfield, Mass.

(389.) — Rupture of a plate on a boiler in the sugar house of O. Richard, Sunshine, La., occurred on October 31st.

(390.) — On October 31st, a boiler exploded in the Jones Meat Market, East Second St., Chillicothe, Ohio. One man was slightly injured.

(391.) — An accident occurred October 31st, to a blowoff pipe in the Light Plant & Water Works of the City of Wynnewood, Wynnewood, Okla.

NOVEMBER, 1919.

(392.) — An accident to a tallow tank occurred November 1st, at the abattoirs of the Retail Butchers Protective Assoc. Company, Cleveland, Ohio.

(393.) — An accident to a boiler on a sand boat of the Schwartz Lumber & Coal Co., Wichita, Kansas, occurred November 1st.

(394.) — On November 2nd, a boiler exploded at the Convent of the Sisters of St. Francis of Assisi, St. Francis, Wis.

(395.) — Three sections of a heating boiler cracked November 3rd, in School No. 8, Sherman Ave., Jersey City, N. J.

(396.) — A tube pulled loose November 3rd, on a boiler on the dredge "Natonia" owned by the port of Astoria, Astoria, Ore. One man was killed and another injured.

(397.) — A tube ruptured in the barrel factory of the Tide Water Oil Company, Bayonne, N. J., November 3rd. Four men were injured.

(398) — A section of a heating boiler cracked November 3rd, in a building used by the Army & Navy Y. M. C. A., City Square, Charleston, Mass.

(399.) — A section cracked November 4th, in a boiler in the Williamstown Opera House, Williamstown, Mass.

(400.) — A section cracked November 4th, in a boiler in the garage at 20 Bowdoin St., Springfield, Mass., owned by H. G. Webster.

(401.) — Three headers in a steam boiler cracked November 4th, at the L. D. S. Hospital, Salt Lake City, Utah.

(402.) — On November 5th, a heating boiler exploded at the Home Garage,

Arma, Kansas. The owner, Mr. E. J. May, was killed and another man was injured.

(403.) — An accident occurred, November 5th, to a blowoff fitting at a plant of the Southern Cotton Oil Co., 456 West 37th St., Chicago, Ill.

(404.) — On November 6th, three men were killed by an explosion of a locomotive boiler at Hewitt's Station, on the Bessemer & Lake Erie Railroad.

(405.) — Five sections cracked, November 11th, in a boiler at the garage of the United Garage & Sales Company, 1766 Stratford Ave., Bridgeport, Conn.

(406.) — A section cracked in a boiler at the plant of H. B. Raffel, 36 East 9th St., New York, on November 11th.

(407.) — On November 13th, an accident to a boiler at Pier C, Port Richmond, Pa., severely injured one man.

(408.) — An accident to a boiler on November 14th, occurred at the Church of the Covenant, Williamsport, Pa.

(409.) — On November 14th, a boiler exploded at the River Falls Normal School, River Falls, Wis.

(410.) — An accident occurred, November 14th, to the main steam pipe in the fibre mill of the Nekoosa Edwards Paper Co., Port Edwards, Wis.

(411.) — One tube blew out and others were damaged at the Sabraton Works of the American Sheet and Tin Plate Company, Morgantown, West Va., on November 15th.

(412.) — Two sections of a boiler were found cracked, November 11th, in a building at 57-66 Public Square, Watertown, N. Y., belonging to the Northern N. Y. Utilities, Inc.

(413.) — On November 17th, at the Botany Worsted Mills, Passaic, N. J., an accident to a boiler caused three men to be injured.

(414.) — On November 17th, a boiler exploded in the basement of the White River Junction House.

(415.) — A section cracked, November 17th, in a boiler at the garage of R. M. Sparks, Nicholasville, Ky.

(416.) — On November 17th, a boiler exploded at the People's Cleaning & Dye Works, 2416 Erskine St., Omaha, Neb.

(417.) — A boiler ruptured, November 17th, at the National Biscuit Company's plant, 7th and Cass Streets, St. Louis, Mo.

(418.) — A section cracked, November 17th, on a boiler in the greenhouse of the Schluaraff Floral Company, Erie, Pa.

(419.) — On November 17th, a tube burst in the boiler room of Henry Sonneborn & Company, Baltimore, Md. Two men were injured.

(420.) — Three men were injured when a tube ruptured in Boiler House No. 1 of the Botany Worsted Mills, Passaic, N. J., on November 17th.

(421.) — A tube blew out, November 18th, in the Finance Building, South Penn. Square, Philadelphia, Pa. Three men were injured.

(422.) — On November 19th, an accident occurred to a steam pipe at Mine No. 2 of the Parra Coal Company, Parra, Ill. Two men were injured.

(423) — A section cracked, November 19th, in a boiler in the office of the Elyria Telephone Company, Elyria, Ohio.

(424) — A tube ruptured, November 19th, in the power station of the Public Service Corporation of New Jersey at Paterson, N. J.

(425.) — An accident to a boiler of the National Veneer Company, West Michigan St., Indianapolis, Ind., occurred on November 20th.

(426.) — On November 21st, a boiler exploded at the mill of G. S. Parmille and Son, Taft, Oregon. Three persons were killed and two injured.

(427.) — On November 21st, a boiler exploded at the Farmer's Gin Plant, Roanoke, Texas. Three men were killed and three injured.

(428.) — An accident occurred, November 21st, to a boiler in the basement of The Harris Garage Corp., 341 Troy Ave., Brooklyn, N. Y.

(429.) — Seven sections of a heating boiler cracked, November 22nd, in the synagogue of the Congregation Ados Israel, Market St., Hartford, Conn.

(430.) — On November 22nd, a heating boiler exploded at the Fawkes Apartment House, 832 Colfax Ave., Denver, Colo.

(431.) — A section cracked in the basement of the Methodist Episcopal Church South, Conway, Ark., on November 22nd.

(432.) — A tube burst, November 22nd, at the plant of the New York Butcher's Dressed Beef Company, 39th St. and 11th Ave., New York City.

(433.) — On November 24th, a locomotive exploded at Macon, Ga., on the Macon & Birmingham Railroad. One man was killed and another injured.

(434.) — A fitting on a blowoff pipe ruptured, November 25th, at the plant of the Southern Cotton Oil Company, 1456 West 37th St., Chicago, Ill.

(435.) — A tube failed, November 26th, at Mine No. 9 of the Old Ben Coal Corporation, West Frankfort, Ill.

(436.) — Two tubes ruptured, November 26th, at the Buckingham Ave. Power House of the Public Service Corporation of N. J., Perth Amboy, N. J.

(437.) — Two sections of a heating boiler in the basement of a theatre at 214 East 14th St., New York, cracked on November 27th.

(438.) — Two sections were found ruptured, November 27th, in the heating plant of the property of The World Realty Company at 15th and Douglas Sts., Omaha, Neb.

(439.) — Two sections cracked, November 27th, in the hardware store of N. E. Blood, Beloit, Kansas.

(440.) — On November 28th, a boiler accident occurred on the steamer President Grant while the vessel was at sea.

(441.) — On November 28th, a boiler exploded at the factory of B. Zelenko and Company, 14 William St., White Plains, N. Y.

(442.) — An accident to a boiler on November 28th, at the bakery of William Seibt, 3155 Cherokee St., St. Louis, Mo., injured one man.

(443.) — A tube pulled out November 28th, in a boiler at the plant of the Lederle Antitoxin Laboratories, Pearl River, N. Y.

(444.) — A section cracked, November 28th, in a boiler belonging to the F. W. Woolworth Company at 461-69 Fifth Avenue, New York City.

(445.) — A header ruptured, November 28th, in a boiler at the Ferguson Building at 319 Third Avenue, Pittsburgh, Pa.

(446.) — A tube ruptured, November 29th, at the plant of The Winchester Repeating Arms Company, New Haven, Conn. One man was injured.

(447.) — On November 29th, a boiler exploded at the sawmill of Hooker & Campen, Alliance, N. C. One man was killed.

(448.) — Three sections cracked, November 29th, at the rope works of the Lambeth Rope Corporation, New Bedford, Mass.

(449.) — Two sections of a heating boiler in the residence of E. B. Chandler, No. 127 West French Place, San Antonio, Texas, cracked on November 30th.

DECEMBER, 1919.

(450.) — Two sections cracked, December 1st, in a boiler at the packing house of Kingan & Company, Jacksonville, Fla.

(451.) — A tube ruptured at the Stimson Lumber Mill, Seattle, Wash., on December 3rd.

(452.) — Two sections cracked, December 4th, at the Whittier School, Bridgeport, Conn.

(453.) — A fire sheet ruptured, December 4th, at the light and pumping plant of the City of Atlantic, Iowa.

(454.) — A section cracked at the lumber mill of Parker & Page Company, East Cambridge, Mass., on December 4th.

(455.) — Three sections cracked, December 4th, at the Ansonia Amusement Company's theater, No. 385 Third Avenue, New York City.

(456.) — A fitting in a blowoff pipe failed, December 4th, at the Waltham Laundry Company, Gifford Ave., Waltham, Mass. One man was injured.

(457.) — A stop valve failed, December 5th, at the Water and Electric Light Plant of North Vernon, Ind.

(458.) — A section cracked, December 5th, at the prevocational school at New Britain, Conn.

(459.) — On December 6th, a boiler exploded in the store of George F. Reid, Grand Rapids, Mich. Five persons, two of them children, were injured.

(460.) — On December 8th, a tube ruptured at the New Marion Power Station of the Public Service Corporation of New Jersey, injuring two men.

(461.) — A tube failed at the factory of the Rhode Island Card Board Company, No. 163 Exchange Street, Pawtucket, R. I., on December 8th.

(462.) — On December 9th, a boiler exploded at the Great Western roundhouse, Des Moines, Iowa. One man was injured.

(463.) — Three sections cracked at the Jewish Hospital, No. 216 East Kentucky Street, Louisville, Ky., on December 9th.

(464.) — Two sections cracked, December 9th, in a boiler belonging to the Thistle Investment Company, No. 501 Highland Avenue, Kansas City, Mo.

(465.) — A furnace on the Tug "David Gibb," belonging to the North State Lumber Company, Charleston, S. C., collapsed on December 9th.

(466.) — One man was injured on December 10th, when a tube failed at the Fayetteville Gas & Electric Plant, Fayetteville, Ark.

(467.) — On December 11th, a boiler exploded on the Collins Brothers farm on the Georgetown Pike, near Cleveland, Tenn. Two men were injured.

(468.) — A tube ruptured in the power plant of the Louisville Street Railway Company, Louisville, Ky., on December 11th.

(469.) — On December 11th, a boiler exploded in the power plant of the C. and E. I. Railway at Danville, Ill. Two men were injured.

(470.) — An acetylene container exploded at the Wisconsin Foundry Company's plant at Madison, Wis., on December 11th. One man was injured.

(471.) — The bottom plate on a boiler blew out, December 11th, at the manufacturing plant of L. R. and Vilas Harsha Manufacturing Company, 401 Lincoln Street, Chicago, Ill.

(472.) — A tube ruptured, December 11th, at the Buckingham Avenue plant of the Public Service Corporation of New Jersey at Perth Amboy, N. J.

The Hartford Steam Boiler Inspection and Insurance Company

ABSTRACT OF STATEMENT, JANUARY 1, 1920.
Capital Stock, . . . $2,000,000.00.

ASSETS.

Cash in offices and banks	$390,221.07
Real Estate	90,000.00
Mortgage and collateral loans	1,426,250.00
Bonds and stocks	5,702,983.62
Premiums in course of collection	597,171.35
Interest accrued	107,590.44
Total assets	$8,314,216.48

LIABILITIES.

Reserve for unearned premiums		$3,715,903.48
Reserve for losses		175,539.16
Reserve for taxes and other contingencies		401,420.50
Capital stock	$2,000,000.00	
Surplus over all liabilities	2,021,353.34	
Surplus to Policy-holders		**$4,021,353.34**
Total liabilities		$8,314,216.48

CHARLES S. BLAKE, President.

FRANCIS B. ALLEN, Vice-President,   W. R. C. CORSON, Secretary.

L. F. MIDDLEBROOK, Assistant Secretary.

E. SIDNEY BERRY, Assistant Secretary.

S. F. JETER, Chief Engineer.

H. E. DART, Supt. Engineering Dept.

F. M. FITCH, Auditor.

J. J. GRAHAM, Supt. of Agencies.

BOARD OF DIRECTORS

ATWOOD COLLINS, President,
Security Trust Co., Hartford, Conn.

LUCIUS F. ROBINSON, Attorney,
Hartford, Conn.

JOHN O. ENDERS, President,
United States Bank, Hartford, Conn.

MORGAN B. BRAINARD,
Vice-Pres. and Treasurer, Ætna Life
Insurance Co., Hartford, Conn.

FRANCIS B. ALLEN, Vice-Pres., The
Hartford Steam Boiler Inspection and
Insurance Company.

CHARLES P. COOLEY,
Hartford, Conn.

FRANCIS T. MAXWELL, President,
The Hockanum Mills Company, Rockville, Conn.

HORACE B. CHENEY, Cheney Brothers
Silk Manufacturers, South Manchester,
Conn.

D. NEWTON BARNEY, Treasurer, The
Hartford Electric Light Co., Hartford,
Conn.

DR. GEORGE C. F. WILLIAMS, President and Treasurer, The Capewell
Horse Nail Co., Hartford, Conn.

JOSEPH R. ENSIGN, President, The
Ensign-Bickford Co., Simsbury, Conn.

EDWARD MILLIGAN, President,
The Phœnix Insurance Co., Hartford,
Conn.

EDWARD B. HATCH, President,
The Johns-Pratt Co., Hartford, Conn.

MORGAN G. BULKELEY, JR.,
Ass't Treas., Ætna Life Ins. Co.,
Hartford, Conn.

CHARLES S. BLAKE, President,
The Hartford Steam Boiler Inspection
and Insurance Co.

Incorporated 1866.

Charter Perpetual.

INSURES AGAINST LOSS FROM DAMAGE TO PROPERTY AND PERSONS, DUE TO BOILER OR FLYWHEEL EXPLOSIONS AND ENGINE BREAKAGE

Department.	Representatives.
ATLANTA, Ga.,	W. M. FRANCIS, Manager
1103-1106 Atlanta Trust Bldg.	C. R. SUMMERS, Chief Inspector.
BALTIMORE, Md.,	LAWFORD & McKIM, General Agents.
13-14-15 Abell Bldg.	JAMES G. REID, Chief Inspector.
BOSTON, Mass.,	WARD I. CORNELL, Manager.
4 Liberty Sq., Cor. Water St.	CHARLES D. NOYES, Chief Inspector.
BRIDGEPORT, Ct.,	W. G. LINEBURGH & SON, General Agents.
404-405 City Savings Bank Bldg.	E. MASON PARRY, Chief Inspector.
CHICAGO, Ill.,	J. F. CRISWELL, Manager.
209 West Jackson B'l'v'd	P. M. MURRAY, Ass't Manager.
	J. P. MORRISON, Chief Inspector.
	J. T. COLEMAN, Ass't Chief Inspector.
CINCINNATI, Ohio,	W. E. GLEASON, Manager.
First National Bank Bldg.	WALTER GERNER, Chief Inspector.
CLEVELAND, Ohio,	H. A. BAUMHART, Manager.
Leader Bldg.	L. T. GREGG, Chief Inspector.
DENVER, Colo.,	J. H. CHESNUTT,
918-920 Gas & Electric Bldg.	Manager and Chief Inspector.
HARTFORD, Conn.,	F. H. KENYON, General Agent.
56 Prospect St.	E. MASON PARRY, Chief Inspector.
NEW ORLEANS, La.,	R. T. BURWELL, Mgr. and Chief Inspector.
308 Canal Bank Bldg.	E. UNSWORTH, Ass't Chief Inspector.
NEW YORK, N. Y.,	C. C. GARDINER, Manager.
100 William St.	JOSEPH H. McNEILL, Chief Inspector.
	A. E. BONNET, Ass't Chief Inspector.
PHILADELPHIA, Pa.,	A. S. WICKHAM, Manager.
142 South Fourth St.	WM. J. FARRAN, Consulting Engineer.
	S. B. ADAMS, Chief Inspector.
PITTSBURGH, PA.,	GEO. S. REYNOLDS, Manager.
1807-8-9-10 Arrott Bldg.	J. A. SNYDER, Chief Inspector.
PORTLAND, Ore.,	McCARGAR, BATES & LIVELY,
306 Yeon Bldg.	General Agents.
	C. B. PADDOCK, Chief Inspector.
SAN FRANCISCO, Cal.,	H. R. MANN & Co., General Agents.
339-341 Sansome St.	J. B. WARNER, Chief Inspector.
ST. LOUIS, Mo.,	C. D. ASHCROFT, Manager.
319 North Fourth St.	EUGENE WEBB, Chief Inspector.
TORONTO, Canada.	H. N. ROBERTS, President Boiler Inspection
Continental Life Bldg.	and Insurance Company of Canada.

It's all In the Firing!

Wasteful Fuel Consumption! High Steam Cost!

Correspondence Course For FIREMEN

A CONSTRUCTIVE SERVICE

extended to anyone interested

in *BOILER ECONOMY and SAFETY*

Write to-day for details to the HOME OFFICE of

THE HARTFORD STEAM BOILER
INSPECTION and INSURANCE CO.

HARTFORD   CONNECTICUT

The Locomotive

DEVOTED TO POWER PLANT PROTECTION

PUBLISHED QUARTERLY

Vol. XXXIII.	HARTFORD, CONN., OCTOBER, 1920.	No. 4

BOILER EXPLOSION AT OWOSSO, MICHIGAN.

Boiler Explosion at Owosso, Michigan.

MANY of the farmers who live in Shiawassee County, Michigan, find an excellent market for their supply of milk at the creamery at Owosso near the center of the county. In addition to the usual creamery business a condensed milk plant is operated there by the Detroit Creamery Company of Detroit, Mich. Some idea of the amount of milk handled at the Owosso creamery may be gained from the statement that about $2000 is paid out per day to the farmers for their product.

An accident to a plant such as this obviously would be of serious consequence to the farmers and to the owners. It was not expected that any accident would happen yet on the 17th of June the business of this creamery was halted — sharply and abruptly. At 5.40 P. M. on the day mentioned a boiler at the plant exploded and the owners were unable to operate or to receive milk from the farmers until provision could be made for the necessary supply of steam to carry on the work.

If the interruption to the work of the creamery had been the only result of the accident there would be little more to tell. But the damage went further than that. The force of the explosion lifted the boiler up in the air and hurled it a distance of 100 feet where it landed, as shown in the illustration on the front cover of this issue of The Locomotive, on the rear of a dwelling house which was moved on its foundation about ten inches. Two other houses on the same side of the street and one on the opposite side were also damaged. A piece of a grate bar was hurled through the window of the house next door and landed on a bed where a child had been sleeping only five minutes before the accident.

The news of the accident spread rapidly and it was not long before a large number of people were on the scene. It was believed that the fireman had been killed and an immediate search of the wrecked boiler room was made but it was not until a half hour after the explosion that his body was found 600 feet away from the creamery. His legs, arms and skull were fractured and death had undoubtedly been instantaneous.

The boiler equipment at this plant consisted of two boilers of the internally-fired, dry-back, Scotch marine type, the one which exploded being known as No. 2. They were 90 inches in diameter and 12 feet, 10 inches long and were built in 1913. The rupture occurred in the furnace which failed as illustrated on the opposite page. Investigation showed that the furnace had collapsed over its entire length

from head to head. The rupture, which started from the center on
top and continued to a point about four inches below the horizontal
center line on each side, was partly by tearing of the plate and partly
by shearing of the rivets. Only one joint of the furnace was affected

VIEW OF COLLAPSED FURNACE.

and, strange to say, the shell of the boiler was not so much as dented.
As will be noted by reference to the illustration, the upper half of
the front course of the furnace was bent until it stood at an angle of
about ninety degrees. At the rear the upper half of the furnace was
pushed down until it lay almost flat against the lower half.

Statements by persons who had been in the boiler room a short
time before the accident and who were otherwise acquainted with the
plant gave evidence that the safety valves had been blown several
times each day and the assistant fireman stated that they had blown

at about 5 P. M. on the day of the explosion. One man made the statement that he had seen about two gauges of water in each boiler just a few minutes before the explosion occurred. Examination of the exploded boiler, however, gave very plain evidence of overheating, although there were no signs of the presence of scale or oil to bring about this condition. It is probable that the water gauge connections became obstructed in some way and a false indication of the water level was given. The cause of the explosion was evidently overheating of the crown-sheet of the furnace which caused it to collapse and then to rupture.

The employees of the creamery were so wrought up by the occurrence that no one thought of the other or No. 1 boiler which still stood in place with a fire in it. The explosion, of course, left this boiler with a free opening for the steam to escape and by the time some one thought to give it attention the water level had fallen so as to cause overheating of the furnace in this boiler also. Its center ring was buckled and its seams so opened that it was unfit for further immediate service.

It was thought at first that it would be some time before the Creamery Company would be able to operate again but fortunately it was able to secure the use of a traction engine which enabled it to receive milk after a day's idleness. This makeshift, however, did not provide for the entire demand for steam and only the separating of the milk and cream could be accomplished. The creamery was forced to await the completion of the new boiler plant before resuming full operation.

In this disaster the Creamery Company found much consolation in its policy with The Hartford Steam Boiler Inspection and Insurance Company under which it was promptly indemnified for the property loss it had sustained and which amounted to approximately $7000.

Forty Years of Boiler Explosions.

A SUMMARY of the records of the past forty years (1880 to 1919 inclusive) shows that there have been in that period a total of 14,281 explosions, 10,638 lives lost and 17,085 persons injured. These figures have been obtained from the explosion lists of THE LOCOMOTIVE and, while we do not claim completeness for these lists, they are fairly representative of conditions from year to year.

On page 102 we have tabulated the data obtained from the above lists under the column headings of the Year, Explosions per Year, Persons Killed, Persons Killed per Explosion, Persons Injured, Persons Injured per Explosion, Total Casualties, Casualties per Explosion and the Ratio of Injured to Killed.

A study of statistics and the changes which have occurred from year to year is always of considerable interest. A tabulation of the figures, however, does not always bring out the salient features at a glance and in order that a better study may be made of the data we are presenting it in the form of charts.

Chart No. 1 below gives the Total Explosions per Year from

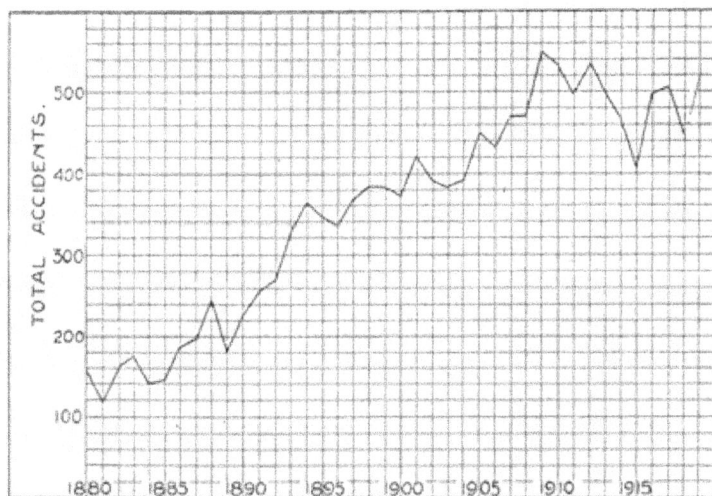

CHART No. 1. — TOTAL EXPLOSIONS PER YEAR.

the year 1880 to 1919 inclusive. Of course the "saw-teeth" are only natural in a chart of this kind. It is interesting to note, however, that

Summary of Boiler Explosions from 1880 to 1919 Inclusive.

Year	Number of Explosions	Persons Killed		Persons Injured		Persons Killed and Injured		Ratio Injured Killed
		Total for Year	Per Explosion	Total for Year	Per Explosion	Total for Year	Per Explosion	
1880	170	259	1.58	555	3.26	814	5.84	.467
1881	119	251	2.11	313	2.63	564	4.74	.801
1882	172	271	1.63	369	2.14	640	3.77	.735
1883	184	263	1.43	412	2.24	675	3.67	.640
1884	152	254	1.67	261	1.72	515	3.39	.972
1885	155	220	1.42	288	1.86	508	3.28	.764
1886	185	254	1.37	314	1.70	568	3.07	.808
1887	198	264	1.33	388	1.96	652	3.29	.688
1888	246	331	1.34	505	2.05	836	3.39	.658
1889	180	304	1.69	433	2.40	737	4.09	.702
1890	226	244	1.08	351	1.55	595	2.63	.695
1891	257	263	1.02	371	1.44	634	2.46	.709
1892	269	298	1.11	442	1.64	740	2.75	.674
1893	316	327	1.03	385	1.22	712	2.25	.850
1894	362	331	.89	472	1.30	803	2.19	.702
1895	355	374	1.05	519	1.46	893	2.51	.720
1896	346	382	1.11	529	1.53	911	2.64	.721
1897	369	398	1.15	528	1.43	926	2.58	.753
1898	383	324	.84	577	1.50	901	2.34	.562
1899	383	298	.78	456	1.19	754	1.97	.654
1900	373	268	.72	520	1.39	788	2.11	.516
1901	423	312	.74	646	1.53	958	2.27	.483
1902	391	304	.78	529	1.35	833	2.13	.575
1903	383	293	.76	522	1.36	815	2.12	.562
1904	391	220	.56	394	1.01	614	1.57	.558
1905	450	383	.85	585	1.30	968	2.15	.654
1906	431	235	.55	467	1.08	702	1.63	.503
1907	471	300	.64	420	.89	720	1.53	.713
1908	470	281	.60	531	1.13	812	1.73	.528
1909	550	227	.41	422	.77	649	1.18	.538
1910	533	280	.53	506	.95	786	1.48	.553
1911	499	222	.44	416	.83	638	1.28	.533
1912	537	278	.52	392	.73	670	1.25	.708
1913	499	180	.36	369	.74	549	1.10	.488
1914	467	148	.37	315	.67	463	1.04	.470
1915	404	132	.33	236	.58	368	.91	.559
1916	499	199	.40	375	.75	574	1.15	.531
1917	506	149	.29	289	.57	438	.86	.513
1918	449	130	.29	340	.76	470	1.05	.382
1919	528	187	.37	343	.68	530	1.05	.543
Total	14,281	10,638		17,085		27,723		
Average			.877		1.38		2.31	2.31

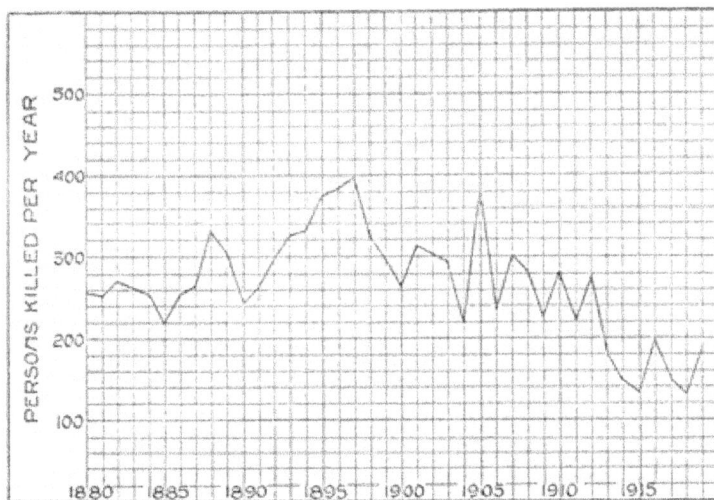

CHART NO. 2. — PERSONS KILLED PER YEAR.

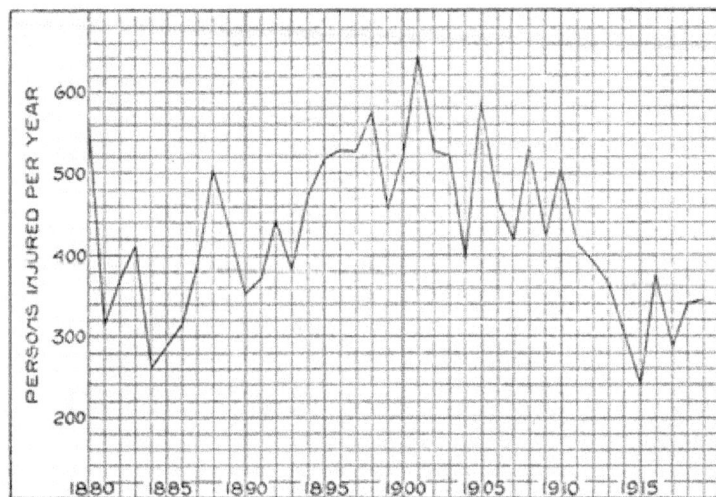

CHART NO. 3. — PERSONS INJURED PER YEAR.

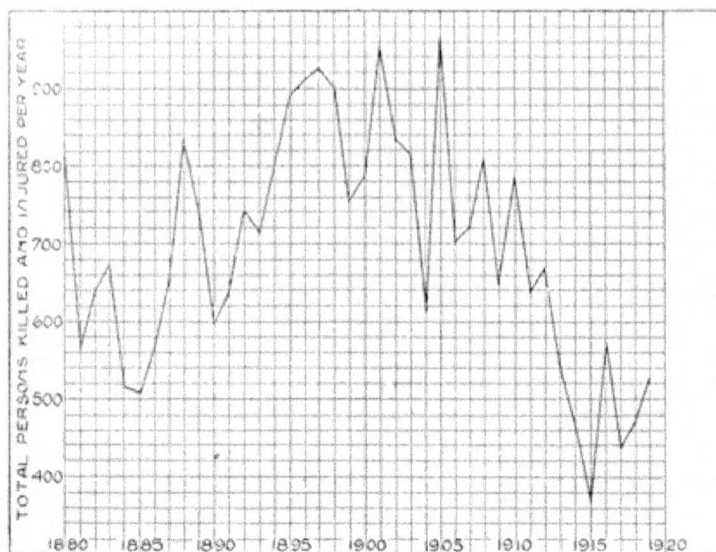

CHART No. 4. — TOTAL PERSONS KILLED AND INJURED PER YEAR.

the average curve rises rapidly until the year 1910 or 1912 is reached,
after which time it is apparently on the decline. It is very probable
that the rise is due to the rapid increase in the number of boilers in
use and on that account the rise should continue to 1919. If accurate
information in regard to the number of boilers in use during each of
these years were available it would be of considerable help but certainly
the number of boilers in use in the past ten years has increased at least
as rapidly as in the previous period. The drop in the curve is undoubtedly due to better construction, better care and more heed to the
advice of the boiler inspector.

Considering next the variation in Persons Killed, as shown in
Chart No. 2, on page 103, we find at first a very slight decrease
followed by a general increase which is interrupted in two successive
years by a decline but which rises again to the peak of 398 in 1897.
A general decrease then begins which again is broken by a sharp upward "saw-tooth" in 1905.

In Chart No. 3 on page 103 we have the variation in Persons
Injured. The irregularities are very marked but there is a general
rise to 646 in 1901 and then a decided falling off in the later years. The

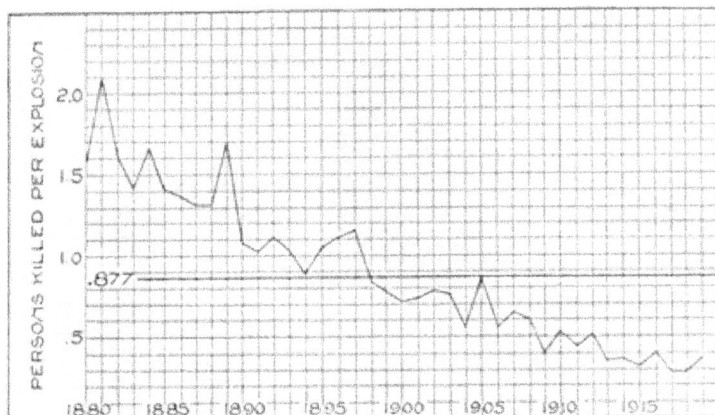

CHART No. 5. — PERSONS KILLED PER EXPLOSION.

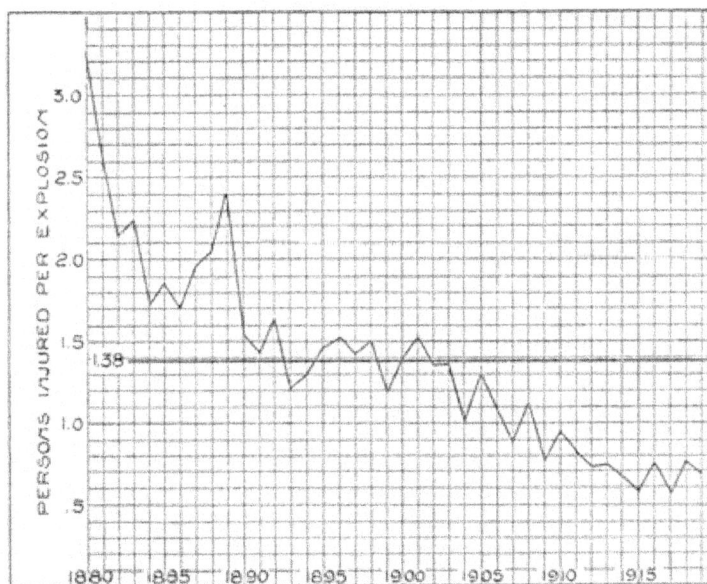

CHART No. 6. — PERSONS INJURED PER EXPLOSION.

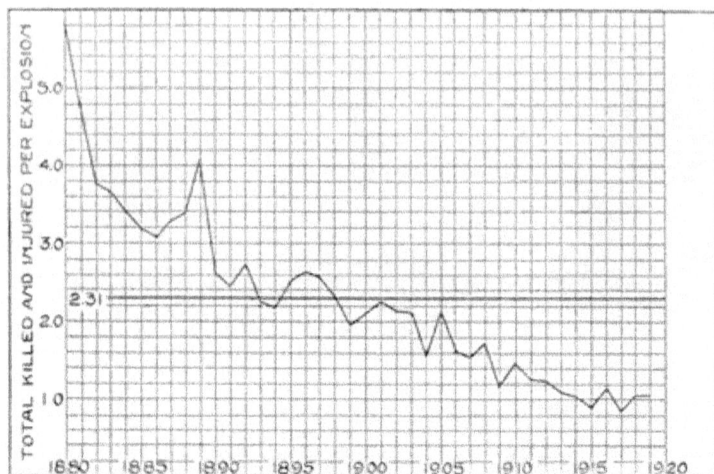

CHART No. 7. — TOTAL KILLED AND INJURED PER EXPLOSION.

average height in the past five years, in fact, is below that in the period from 1880 to 1885.

The variation in the Total of Killed and Injured as shown in Chart No. 4 on page 104 and which is, of course, the result of the addition of the figures of Charts Nos. 2 (Persons Killed) and 3 (Persons Injured), shows very markedly the effect of the " saw-teeth " of Chart No. 3. Here again, however, there is a general rise until the peak is reached, in this case in 1905, followed by a decline. Also, as in Charts Nos. 1 and 3, we find the average in the latter years to be lower than in the earliest years.

While all of the above charts show the facts they do not indicate the essential points as regards the casualties. To get a better conception of what has happened we must consider not the Persons Killed, Persons Injured and Total of Killed and Injured but rather the rate of each of these per explosion. These figures are given in Columns No. 4, 6, and 8 respectively. In each of these columns the figures have been obtained by dividing the respective casualties by the explosions for the year in question. The graphic analyses of these last three sets of figures are given in Charts No. 5 (Persons Killed per Explosion), No. 6 (Persons Injured per Explosion) and No. 7 (Total of Killed and Injured per Explosion) on pages 105 and 106. In every case we find a most marked falling off in the

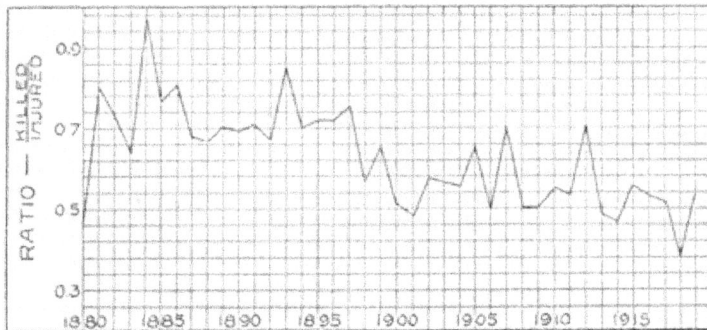

CHART No. 8. — RATIO $\frac{\text{KILLED}}{\text{INJURED}}$

rate. All are broken by "saw-teeth" yet at the same time they are remarkably smooth for charts of data of this nature.

Referring to Chart No. 5. that of the Persons Killed per Explosion, we find a horizontal line drawn at the height marked .877. This value is the average yearly rate of persons killed per explosion and is the result of dividing the sum of all the yearly rates by the number of years (40). This is not the same figure that we would obtain by dividing the total killed during the forty years by the total number of explosions during that period. The former figure is the average yearly rate whereas the latter figure is the rate over a period of forty years. For the forty year rate. in the case of persons killed per accident, we would have 10,637 divided by 14,259 or .747. If the variation followed a straight line of uniform slope or inclination these two rates would be identical.

Figures for average rates similar to that just described on the rate of Persons Killed per Explosion are given on the charts for Persons Injured per Explosion (Chart No. 6) and Total of Killed and Injured per Explosion (Chart No. 7) and are plotted as 1.38 and 2.31 respectively.

Chart No. 8 above represents the variation in the Ratio of Killed to Injured. It shows a slight though definite and rather regular decline. Reference to this chart will show that in the earlier years there were almost as many killed as there were injured whereas at the present time only about one half as many persons are killed as there are injured.

Charts No. 1 (Total Explosions per Year), No. 5 (Persons Killed per Explosion), No. 6 (Persons Injured per Explosion), No. 7 (Total of Killed and Injured per Explosion and No. 8 (Ratio of Killed to Injured) are important and interesting in their indications. The statistics of the last ten years furnish ample evidence that more attention is being given to safe construction and careful operation than in former years. Fewer explosions per number of boilers installed, lower rates of casualties per explosion and a lower ratio between killed and injured are all movements in a fortunate direction. But only untiring vigilance will keep them moving in that way.

Vulcanizers and De-Vulcanizers.

H. E. Dart, Superintendent of Engineering Department.

SINCE the process of vulcanization was patented by Charles Goodyear in 1844 pure india-rubber or caoutchouc has been very rarely used, the vulcanized rubber being far preferable for almost every use. In general, the process of vulcanization consists in mechanically mixing rubber at moderate heat with sulphur and then "curing" it in steam at from 250 to 300 degrees, Fahrenheit. The substance thus formed acquires extraordinary powers of resisting compression together with a great increase of strength and elasticity; it remains elastic at all temperatures, it cannot be dissolved by ordinary solvents, and it is not affected by heat within a considerable range of temperature. Various modifications of the process are employed as, for example, the mixture with the rubber and sulphur of other ingredients such as litharge, white lead, zinc-white, whiting, etc., giving color, softness and other qualities to the product. By using a greater proportion of sulphur and a higher temperature, hard rubber, known as vulcanite or ebonite, can be produced.

Since the process of vulcanization was perfected the uses of rubber have multiplied so that it is now employed in almost every department of industry. Some of the things for which it is used include tires, boots, shoes, mats, toys, belting, buffers, water-proof cloth, washers, valves, fire-hose, medical and surgical appliances, life preservers, steam and water packing, tubing, artificial sponges, etc.

Vulcanizing on a large scale is usually done in a long horizontal cylinder having one head hinged so that it can swing outward and thus provide access to the interior of the vessel by means of an opening of the full shell diameter. Rails are placed along the bottom of the cylinder and the articles to be vulcanized are placed on small cars running on these rails, provision thereby being made for ease in handling For some special **work** jacketed vulcanizers **are used**. These vessels are of the same general form as the ordinary vulcanizer except that they have two concentric shells, the steam being admitted to the space between the shells so that the work is not exposed to moisture. Of course the inner shell is subject to collapse due to external **pressure** and it is therefore supported from the outer shell by means of staybolts in the same manner as some types of boilers. In some cases a vacuum is used in the inside shell, thus increasing the tendency to collapse, but there is no difficulty in designing the vessel so as to provide against this contingency. Small vulcanizers are sometimes placed in a vertical position with the lower head fixed and the entire shell removable by a chain-block or other similar means.

Devulcanizers are similar to vulcanizers in construction but the temperature (and consequently, the pressure) employed is higher than for vulcanizing.

Vessels of this general type are subject to a hazard not found in other kinds of pressure vessels, due to the necessity of providing such a large cover which must be amply strong and at the same time capable of quick opening and closing, and the failure of such a vessel is liable to be very destructive and is frequently accompanied by loss of life. The cover is generally attached to the shell by means of two flanged rings, one of which is riveted to the open end of the vulcanizer and the other to the cover. The two rings are hinged together at one side and a wheel roller, traveling on the floor, is provided at the bottom of the covering to take the weight off the hinge. Eye-bolts are attached to the vulcanizer ring by means of brackets cast on the back side of the ring and these bolts are designed to swing into slots in the cover ring when the cover is closed; the cover is then secured in position by tightening up nuts on the ends of the eye-bolts. Of course packing is necessary to make a steam-tight joint and this is placed in a groove turned in the vulcanizer ring. On account of their size and shape the rings are cast, thus introducing the usual uncertainty as to the quality of the metal. Cast steel should always be used for the purpose instead of cast iron. There is no simple formula for calculating the dimensions of these rings; in fact, the problem is considerably complicated and the article which appeared in The Locomotive for July, 1905, is probably as good as anything that has been published upon the subject. Several patented styles of quick-closing vulcanizer doors are on the market but not all of these are acceptable for insurance.

As regards the proportions of vulcanizers, a diameter of 60 inches is not uncommon; our Engineering Department has designed some with a length of over 50 feet but 16 to 20 feet is more usual. For a pressure of 125 pounds the head of a 60-inch vulcanizer would carry a load in excess of 175 tons and would require twenty-four 2¼-inch cover-bolts. Instead of tightening the bolts up gradually all around the circumference so that they will just hold the pressure, ignorant or careless workmen will often strain the bolts by using a long wrench with a pipe over the end. In this way it is possible to strip the threads after awhile, especially since the nuts must fit loosely as a matter of convenience in the repeated handling to which they are subjected. An added hazard sometimes is introduced when the workmen do not tighten up all of the bolts — either from laziness on the part of the men or because some of the bolts may be out of order.

The inherent hazard that exists in this class of apparatus will be illustrated by the following partial list of vulcanizer failures taken from the reports given in The Locomotive: —

May 6, 1912. Vulcanizer exploded at the Plymouth (Mass.) plant of the Boston Woven Hose and Rubber Co. Property loss said to have been about $10,000.

May 16, 1912. Vulcanizer exploded at plant of Empire Rubber Co., Trenton, N. J. One man killed and one fatally injured (Illustrated in The Locomotive of October, 1912.)

Jan. 21, 1914. Vulcanizer exploded in garage at Nashua, N. H. Large property damage.

June 9, 1916. Door of vulcanizer blew off at plant of Electric Hose and Rubber Co., Wilmington, Del. One killed and three injured.

Feb. 10, 1917. Vulcanizer exploded at plant of Luzerne Rubber Co., Trenton, N. J.

Jan. 17, 1919. Vulcanizer exploded at the tire shop of Samuel Bros., Philadelphia, Pa. Two men were seriously injured.

Feb. 10, 1920. Vulcanizer exploded at Durham, N. C. One killed and one injured. See explosion list in this issue, page 125, No. 106.

Supersaturated Steam.

THE terms "wet steam," "dry or saturated steam" and "super-heated steam" are so familiar to every one who has but a slight acquaintance with water and its vapor that it would seem unnecessary to say anything further in regard to them. But now that we are hearing a good deal of the term "supersaturated steam," indicating, as it does, a condition of vapor somewhat unique, it may be well, before we discuss this condition of steam, to hark back to some first principles.

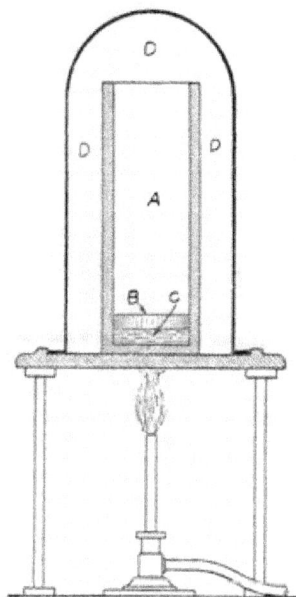

FIG. 1.

Let us suppose that we have an apparatus as shown in fig. 1 which consists of a cylinder A which has a piston B of an ideal nature such that it fits the cylinder perfectly and yet will move with absolutely no friction. Its weight is such that it will exert a pressure of 14.7 pounds on each square inch of the surface of the water C. We will suppose also that the water weighs exactly one pound. All the air has been exhausted from the surrounding space D so that the only pressure that can be exerted on the water will be 14.7 lbs. per square inch. By the nature of the arrangement the pressure will always be constant at 14.7 lbs. per square inch and cannot change as long as the piston is within the cylinder A. We will assume further that we have some means of supplying heat to the water as indicated in the sketch. If we start with the water at the temperature of 32° Fahrenheit and add heat to it the water will increase very slightly in volume but will still remain a liquid until the temperature of 212° is reached. At this point, if we continue to supply heat, the temperature will remain constant at 212° and the water will begin to change from a liquid to a vapor. It will continue to follow this change with a very rapid increase of volume at a constant temperature until the water is all vaporized. When and just when the vaporization is complete the volume occupied by the vapor will be 26.79 cubic feet. Of course, to do this, our apparatus would necessarily have to be of somewhat different proportions than indicated in the sketch. We will now have in our cylinder one pound of dry saturated vapor. We will have in that space occupied by the vapor a definite number of molecules or particles of extremely minute size although we shall not attempt to state the *exact* number.

If we continue to add heat to the vapor it will continue to increase in volume and it now also will undergo a temperature rise. It is no longer "dry saturated steam" but "superheated steam" and if we raise the temperature of it to 262° F. it will be 50° higher than when it was "dry saturated" and we would speak of it as being "steam at 14.7 lbs. and 50° superheat." Since the total number of molecules has not changed but the volume has increased it necessarily must be true that each cubic inch of the superheated vapor contains a smaller number of molecules than did a cubic inch of the saturated steam.

If now we cool the steam it will return to the dry saturated condition and if we cool it still further it will begin to condense and this condensation will continue until it returns to its initial condition of a liquid. In the condition in which it exists between dry saturated steam and a liquid it is known as wet steam or, more correctly, as wet saturated steam. This latter term, that of wet saturated, might appeal to us as incorrect. Actually, however, it is entirely proper. The steam in this condition consists of a mixture of a vapor and a liquid. The liquid may be present as a spray or mist or it may exist as one body of water. The vapor part, however, is still a dry saturated vapor. Each cubic inch of the *vapor* contains just as many molecules as it did when the entire pound was "dry saturated." This term, saturated, is a very good one. It means that we have put into a cubic inch of the vapor as many molecules as can conveniently be placed in that volume. In other words, we have *saturated* the cubic inch with water vapor molecules.

Using the same apparatus we might have had a piston that would have exerted a pressure of 100 pounds per square inch. We would have found, under this condition, that the water could not exist as dry saturated steam, until the temperature was raised to 327.8° F. For this steam to be superheated 50° the temperature would have to be 377.8° F. And in the same way we would find that for every other pressure that we might impose there would be a corresponding temperature for the dry saturated steam.

It is important to note that the temperature of the dry saturated steam and of the wet saturated steam is the same. The vapor and

the liquid are both at the same temperature and, provided no heat is
removed from or added to the mixture, they will remain indefinitely
in the proportions in which they are existing. If 75% of the pound
is vapor and 25% is liquid then the mixture is known as 75 per cent.
"quality" steam. This condition of both vapor and liquid being at
the same temperature with no tendency to change in relative proportion is known as "thermal equilibrium." There is no more tendency
for heat to leave the vapor and enter the liquid and thereby vaporize
that liquid, than there is for heat to pass from the liquid to the vapor.

As a further illustration of what is meant by thermal equilibrium
we may consider that we have a vessel filled with vapor at a given
temperature and pressure. If we introduce some water, the temperature of which is lower than that of the vapor, into this vessel some
of the vapor will condense until it has given up enough heat to raise
the temperature of the liquid to that of the vapor. If the water is
at a temperature higher than the vapor then some of the water will
vaporize until the temperature of both becomes the same.

This idea of thermal equilibrium plays an important part in supersaturated steam. It might be said at this point that the condition of
supersaturated steam is an unusual one in that it is a very unstable
condition. While scientists long have known of the possibility of its
existence it has not generally been supposed that it would be found
in practice to any appreciable extent.

In order to appreciate this condition of supersaturation it must
be fully realized that when we have thermal equilibrium the water
vapor at any pressure must have a perfectly definite and easily determined temperature. Furthermore, if it is dry saturated, it will occupy
a certain definite volume known as the *Specific Volume*. If it is 50%
wet it will occupy very nearly 50% of the Specific Volume.

In investigating the conditions under which steam turbines operate
it has been found repeatedly that if a pressure gauge and a thermometer are used to read the pressure and temperature of the steam
in the exhaust passage the thermometer will read a temperature
several degrees below the value corresponding to the pressure. If
thermal equilibrium existed no such discrepancy in readings could
exist. The condition has sometimes been spoken of as "undercooling" and this term is a very good one. As a matter of fact the
vapor is lower in temperature than it would be under equilibrium
conditions for the observed pressure. And yet it is a difficult matter
to obtain the exact temperature of the vapor. The thermometer, if
placed in a pocket where it will be protected from the rush of the

steam, will soon be covered with a film of moisture due to condensation of the vapor. Condensation of the supersaturated vapor under these conditions will always occur because, as has been said before, this unique condition is a very unstable one. The vapor will always tend to return to a condition of thermal equilibrium with the condensate and the thermometer mentioned above will therefore read very nearly or exactly the temperature corresponding to the observed pressure. If, however, the thermometer bulb is exposed to the high velocity steam the film of moisture will be swept away and the molecules of vapor which are at the lower temperature will be given an opportunity to register their true value. Even this method of taking the temperature must give only approximate results

While the above discussion of " undercooling " serves to bring out some remarkable facts concerning supersaturated steam it does not give us a conception of just what is meant by the term " supersaturation." An attempt will now be made to explain this term.

The conditions under which steam flows through a nozzle and the theoretical laws which govern this flow have long been known. These laws had always been considered very exact and absolute. It was thought that the actual flow of steam through a nozzle would have to be less than the theoretical on account of friction impeding the flow. When, however, tests showed that the actual flow of the practically dry steam used in the tests was *greater* than the theoretical it was realized that something unusual had occurred. At first it was thought that moisture in the steam was the explanation of the greater weight discharged. But when calculation determined that from 10% to 20% of moisture would have to be present to give the results obtained it was realized that some other explanation must be secured. Since the weight discharged must be in direct relation to the number of molecules passing in a given time the only possible solution is that the vapor must carry more molecules per cubic inch under the actual conditions of flow than when the unit volume is " saturated " with molecules. In other words, the unit volume is " supersaturated." As an actual fact we do not know how many molecules it would be possible to crowd into a unit volume but we do know that only a certain definite number will stay there naturally and with this number in the unit volume the steam is saturated. Note that this is the only natural condition in which *vapor* will exist. While it may become supersaturated it will very readily and upon the slightest disturbance revert to the saturated state with consequent condensa-

tion of its excess molecules and when this change is effected thermal
equilibrium will be established.

The subject of supersaturation is one that is demanding very close
study by the steam turbine designer. The matter has been the more
forcibly brought to the attention of engineers and scientists in this
field by an article entitled "A New Theory of the Steam Turbine"
by H. M. Martin in Engineering (London), Vol. CVI. Mr. Martin
brings out the point that reciprocating engine design has been largely
by cut and try methods with theoretical considerations giving aid only
in a most general manner. The varying conditions in different engines
such as temperature changes within each cycle of operation, valve
leakage and clearance have been most difficult of correlation between
any given engine and a new design. It is true that theory has been
applied wherever and whenever possible but the gap between theory
and practice has been a wide one. The steam turbine, however, has
lent itself most favorably to theoretical design. Predictions of the
performance of a given design have been possible to a remarkable
exactness. Even here considerable experience has been necessary to
determine certain correcting factors for discrepancies between theory
and practice. Mr. Martin suggests that some of these apparent discrepancies may be explained by the fact that the steam is super-
satured during all, or very nearly all, of its passage from the nozzle to
the condenser.

The variable quantities encountered in an investigation of this
nature are many. It would seem almost impossible to develop formulæ
for use in design. Similar obstacles have arisen in the past, however,
and sooner or later have been surmounted. It is therefore confidently
expected that the investigators in this field will reach a complete and
satisfactory solution of the whole problem.

In these days of high prices and crippled transportation the question of saving coal has become more important than ever before. The
firemen are those who are in a position to effect a saving. Are your
firemen equipped with the information to enable them to fire efficiently?
The Hartford Correspondence Course in Boiler Economy and Safety
is proving itself of immense benefit to firemen in many sections of the
country. If you are interested and would like further information write
The Correspondence Course Department, Hartford Steam Boiler Inspection & Insurance Co., Hartford, Conn.

The Locomotive

Devoted to Power Plant Protection

Published Quarterly

Wm. D. Halsey, Editor.

HARTFORD, OCTOBER, 1920.

Single copies *can be obtained free by calling at any of the company's agencies.*
Subscription price 50 cents per year when mailed from this office.
Recent bound volumes one dollar each. Earlier ones two dollars.
Reprinting matter from this paper is permitted if credited to
The Locomotive of the Hartford Steam Boiler I. & I. Co.

WE desire to call the attention of our readers to the leading article in this number of THE LOCOMOTIVE as indicating the trend of boiler accidents and the deaths and injuries which such accidents leave in their wake.

Possibly our treatment of the subject may seem cold blooded. Statistical study usually is of that nature. But we have no desire to leave that impression on our readers. Indeed our sympathies should be the more deeply touched when we think of the suffering a disastrous boiler explosion can bring. It is true that conditions are far better than they were in former years yet when we consider that during the past year there were 528 boiler explosions, 187 persons killed and 343 persons injured, we have cause to ask for the reason and the remedy.

The boiler manufacturer of today is extremely earnest in his efforts to produce not only an efficient boiler but one that is well constructed and safe in every detail. A very great degree of success has been obtained in this direction. But no boiler ever made was intended to be abused. It is also true that no one willfully would subject a pressure vessel, capable of doing such extreme damage, to carelessness and neglect. Unfortunately, however, mankind is not infallible. As long as there are boilers in operation; as long as these boilers must be subjected to the failings of the human race; just so long will we have boiler explosions and failures.

How many of the people of this country, other than boiler inspectors, would be willing after an inspection of a boiler to certify that it was safe to operate? Certainly only a small fraction of one per cent. Boiler inspection is not a thing to be learned in a day. It takes years of experience to learn the danger signals. The purpose of inspection is to warn of trouble and the purpose of insurance is to cover financial losses. Are YOU protecting your equipment, your employees, and your bank account?

We are always pleased to note in the pages of our contemporaries any reference to the value of boiler inspection and we therefore have read with considerable interest two recent articles of this nature.

In an article in the July issue of *Safety Engineering*, entitled " Value of Steam Boiler Inspections in Mining and Industrial Plants " by Mr. H. M. Motherwell of the United States Bureau of Mines, the imperative need for boiler inspection is brought out with exceeding clearness. Some interesting figures also are given on the economic losses attending the neglect of boilers.

Mr. W. E. Snyder of the American Steel and Wire Company in an article in *Power Notes*, published by the Diamond Power Specialty Co., lays great stress on the value of Safety First in the boiler room. Speaking of boiler inspection he says, " The benefits which result from periodic inspections by men whose sole time is taken up with such work, and who in consequence acquire an experience which makes them very keen detectors of defects, are so great that they cannot be over-emphasized."

Personal.

Mr. John T. Coleman, Assistant Chief Inspector of our Chicago Department, has been transferred from the main office of that department in Chicago to its branch in Detroit in order that he may have more direct supervision of the Company's inspection service in the territory adjacent to the latter city. The business prosperity of that automobile center is reflected in the increased demand there for our inspection and insurance protection, which service Mr. Coleman's transfer will enable us to afford more promptly and efficiently than heretofore. We are confident that our policy holders in that section will appreciate the advantage of this move.

Mr. Charles W. Zimmer, who has acquired proficiency in Hartford inspection methods and requirements by his service of over twenty years as an inspector of the Company has been promoted to the position of Assistant Chief Inspector and will fill the vacancy in Chicago created by Mr. Coleman's transfer. The ability and zeal for the Company's interests which Mr. Zimmer has shown in his work are thus rewarded in a manner which we believe will be gratifying to the many friends he has made among the patrons of our Company.

Boiler Explosions.

(INCLUDING FRACTURES AND RUPTURES OF PRESSURE VESSELS)
DECEMBER, 1919, continued.

(473.) — On December 11th a heating boiler exploded at the Home Garage, Arma, Kansas, owned by Mr. Edw. J. May. Mr. May was killed and one other man was injured.

(474.) — A section of a heating boiler cracked at the factory of H. B. Raffel, 36 East Ninth Street, New York City, on December 11th.

(475.) — On December 12th, a boiler exploded at the Home for Aged Women, Cedar Rapids, Iowa.

(476.) — A pipe on the water column of a boiler in the Congregational Church at South Deerfield, Mass., failed on December 12th.

(477.) — One man was killed at the Sioux City, Iowa, Gas & Electric plant when a tube ruptured on December 12th.

(478.) — A tube ruptured, December 13th, at Plant No. 1, Huston Coal & Coke Company, Elkhom, West Va.

(479.) — A heating boiler accident occurred on December 13th, at the Brass Foundry Company, 711 South Adams Street, Chicago, Ill. One man was injured.

(480.) — On December 13th, a boiler exploded at the nursery of E. Enomoto, Redwood City, Calif. One man was injured.

(481.) — Six sections of a heating boiler cracked, December 13th, in the Jefferson School, Butte, Montana.

(482.) — Accidents to the heating boiler in the Simpson Memorial Church, Long Branch, N. J. on December 14th, necessitated the installation of a new boiler.

(483.) — A hot water tank exploded on December 14th, at the home of Mrs. Margaret Kelley, 664 St. Paul Street, Memphis, Tenn. Mrs. Kelley was severely injured.

(484.) — One tube was ruptured and nine others were damaged at the College of Wooster, Wooster, Ohio, on December 15th.

(485.) — An acetylene tank exploded, December 15th, at the Remmert Manufacturing Company's plant, Belleville, Ill. One man was injured.

(486.) — The head of a diffusion tank blew off on December 15th, at the factory of The Chesapeake Pulp & Paper Company, West Point, Va.

(487.) — A tube ruptured, December 15th, in the Union Stock Yards, Section 44, of Armour and Company, Chicago, Ill.

(488.) — A boiler sheet ruptured, December 15th, at the Sibley Electric Light Plant, Sibley, Iowa.

(489.) — Two sections of a heating boiler belonging to Asa C. Isham, Main and Maple Avenues, Norwood, Ohio, cracked on December 15th.

(490.) — On December 15th, a boiler exploded at the Blickham Ice House, Quincy, Ill.

(491.) — On December 16th, the boiler of a locomotive exploded on the B. & O. Railroad at Newcastle, Pa. One man was killed and two were injured.

(492.) — One man and six children were injured on December 16th, when a boiler in the High School at Mountain City, Tenn., exploded.

(493.) — On December 16th, a boiler exploded at the Municipal Golf Links, Seattle, Wash.

(494.) — A section of a heating boiler cracked December 16th, in the office building of W. A. Davis, 101 West Jefferson Street, Syracuse, N. Y.

(495.) — Three sections of a heating boiler in the hotel at 2139-45 Washington Street, Roxbury, Mass., owned by James E. Doherty & Company, cracked on December 17th.

(496.) — A steam separator ruptured, December 17th, at the Rogers-Brown Iron Company, Buffalo, N. Y.

(497.) — A section of a heating boiler cracked, December 17th, at a building at 16 East 18th Street, New York City, owned by Najeeh & Phillip Kaime.

(498.) — Two sections of a heating boiler in the Pittsburgh Plate Glass Company's building at 2406 Albion Street, Toledo, Ohio, were cracked on December 17th.

(499.) — On December 18th, an accident to a heating boiler in the basement at 1009 Arch Street, Philadelphia, Pa., badly damaged the heating plant.

(500.) — Two headers of a boiler cracked, December 18th, in the power plant of the Austin Street Railway Company, Austin, Texas.

(501.) — Five sections of a heating boiler cracked in the apartment house owned by Miss M. W. Knapp, at 96 Plymouth Avenue, Rochester, N. Y., on December 18th.

(502.) — On December 19th, an accident occurred to the heating plant of the Bay View Bottling Company, 11 Hamlin Street, South Boston, Mass.

(503.) — On December 19th, a boiler exploded in the basement of the Penn Beef Company's store, 2550 Germantown Avenue, Philadelphia, Pa.

(504.) — A cast steel fitting at the flour mill of the Marshall Milling Company, Marshall, Minn., ruptured on December 19th.

(505.) — A steam pipe fractured, December 19th, at the mill of the Buckeye Cotton Oil Company, Jackson, Miss.

(506.) — Two sections of a heating boiler cracked December 19th, in the apartment house at 1261-65 Main Street, Hartford, Conn., owned by Saul Berman and Benjamin B. Cion.

(507.) — The head on a boiler at the Valley Paper Company, Holyoke, Mass., cracked on December 19th.

(508.) — A tube ruptured, December 19th, at the plant of the MacSimbar Paper Company, Otsego, Mich.

(509.) — A section of a heating boiler cracked, December 19th, at the plant of the Gotham Can Company, 60 Eagle Street, Brooklyn, N. Y.

(510.) — A section of a heating boiler in the apartment house of Theodore and William D. Dellert, 44 Myrtle Street, Pittsfield, Mass., cracked on December 20th.

(511.) — The furnace of a logging boiler belonging to George Newell, Eureka, California, collapsed on December 20th.

(512.) — A tube ruptured, December 21, at Mill No. 1 of the Monroe Binder Board Company, Monroe, Mich.

(513.) — Two headers in a boiler at the chemical plant of the Diamond Alkali Company, Fairport, Ohio, cracked on December 21st.

(514.) — Two sections of a heating boiler cracked, December 22nd, at the Columbus Sanitarium, Columbus, Ohio.

(515.) — A tube was ruptured and a header fractured at Plant No. 2 of the Ford Motor Company, Highland Park, Mich., on December 22nd.

(516.) — A fire sheet ruptured, December 22nd, in the greenhouse of the E. G. Hill Company, Richmond, Ind.

(517.) — Two sections of a heating boiler cracked in the store and loft building at 73 Canal Street, New York City, owned by Ephriam Siff, on December 24th.

(518.) — A blowoff line failed, December 26th, at the plant of the Erie Dyeing Company, 1842 East 40th Street, Cleveland, Ohio.

(519.) — On December 26th, the boiler of a locomotive on the Canadian Northern R. R. exploded at Bartlett, Minn. Three men were injured.

(520.) — A section of a heating boiler ruptured, December 26th, at the garage of the Service Motor Truck Company, 2617-25 South Wabash Avenue, Chicago, Ill.

(521.) — A blowoff pipe failed at the Red Oak Electric Company's plant, Red Oak, Iowa, on December 26th.

(522.) — The drum of a boiler was ruptured, December 27th, at Washer No. 2 of the Southern Cotton Oil Company, Fort Meade, Fla.

(523.) — A tube ruptured on December 27th, at the No. 1 Works of the Jamison Coal & Coke Company, at Luxor, Pa. Three men were injured.

(524.) — On December 29th, a steam pipe ruptured at the mill of the Mardez Lumber Company, Benford, Texas. Two men were killed.

(525.) — On December 30th, an ammonia tank exploded at the plant of the Fulton Ice & Coal Company, Atlanta, Ga. Three men were injured.

(526.) — Four sections of a heating boiler cracked in the store building of Arthur N. Whittemore, 60 Arlington Street, Worcester, Mass., on December 30th.

(527.) — A tube ruptured, December 31st, at the Cadogan Mine of the Allegheny River Mining Company, at Nicholson Run, Pa.

(528.) — A tube ruptured at the Butler Street Power Plant of the Georgia Railway & Power Company, Atlanta, Ga., on December 31st.

JANUARY, 1920.

(1) — Three men were killed and three others seriously injured when a locomotive exploded, January 1st, on the Western Maryland Railroad at Clear Spring, Md.

(2.) — A boiler was damaged by an accident which occurred January 1st, on a log loader belonging to the Hill City Railway Company, at Hill City, Minn.

(3.) — A section of a heating boiler cracked, January 1st, in the Hartford School, Canton, Ohio.

(4.) — On January 2nd, a hot water heater exploded at the home of Leslie Miller, Cedarville, N. J.

(5.) — A flue ruptured, January 2nd, at the plant of the Jacob Dold Packing Company, Wichita, Kansas.

(6.) — One section of a heating boiler cracked, January 2nd, at the Pomfret School, Pomfret, Conn.

(7.) — One man was fatally injured when a tube burst, January 2nd, at the power plant of the Los Angeles Gas and Electric Company, Los Angeles, Calif.

(8.) — A tube failed, January 2nd, at the plant of the American Sheet and Tin Plate Company at New Castle, Penn.

(9.) — Two sections were cracked January 2nd, in the schoolhouse at 8th and Pine Streets, Traverse, Mich.

(10.) — A furnace sheet cracked in a girthwise direction in a boiler at the box factory of the McFerson & Foster Company, Evansville, Ind., on January 3rd.

(11.) — A heating boiler exploded January 4th, at the home of Dr. F. H. Lord of Plano, Ill. Mrs. Lord was fatally injured.

(12.) — Two sections cracked January 4th, in the basement of the United Supply Company store at Gary, W. Va.

(13.) — One man was injured when a tube ruptured January 4th, at the West End Gas Works of the Public Service Corporation, St. Pauls and Duffield Ave., Jersey City, N. J.

(14.) — Six men were injured when a crown sheet ruptured on January 4th at the plant of the Sinclair Gulf Oil Company, Hominy, Okla.

(15.) — A section cracked January 4th, in the apartment house at 1650 California St., San Francisco, Calif., belonging to L. Palmer and M. Stevens.

(16.) — Several tubes were damaged in an accident at the Carnegie Steel Company's plant at Upper Union Mills, Youngstown, Ohio, on January 4th.

(17.) — Accident to a heating boiler at the Graphic Theater, Bangor, Maine, on January 5th, necessitated closing the place until repairs could be made.

(18.) — A tube failed on January 5th, at the Gorge Power Plant of the Northern Ohio Traction and Light Company, Akron, Ohio.

(19.) — A hot water heater cracked January 5th, in the basement of the property of Charles Hirschhorn, 2541 Broadway, New York City.

(20.) — A tube ruptured January 6th, at Plant No. 2 of the Savage Arms Company, Sharon, Penn. Two men were injured.

(21.) — A section of a heating boiler was ruptured January 6th, at the East Ward School, Hastings, Neb.

(22.) — On January 6th, a heating boiler exploded in a garage near the Manhattan bridge on Madison St., New York City. Two men were injured.

(23.) — Two tubes failed January 6th, in the office building of A. R. Buell, Lorain, Ohio.

(24.) — A section of a heating boiler cracked January 6th, in the packing house of Kingan and Company at Jacksonville, Fla.

(25.) — Two tubes failed at the factory of the Wescott Motor Car Company, Springfield, Ohio, on January 6th.

(26.) — A collar brace failed January 6th, in the Ottawa Street Station of the City of Lansing, Lansing, Mich.

(27.) — Three sections of a heating boiler cracked at the Prevocational School, New Britain, Conn., on January 7th.

(28.) — On January 8th, a locomotive exploded at the Chesapeake and Ohio yards at Silver Grove, Ky. One man was injured.

(29.) — A rupture occurred in a boiler at the power plant of the Brush Light and Power Company, Brush, Colo., on January 8th.

(30.) — A blowoff pipe ruptured January 8th, in the factory of the F. J. Lewis Mfg. Company, 130 Second St., Moline, Ill.

(31.) — A tube ruptured January 8th, at the factory of the Kansas City Packing Box Factory, Adams and Wyoming Street, Kansas City, Kan.

(32.) — Four sections of a heating boiler at the New Grade School No. 2, Chassell, Mich., cracked on January 9th.

(33.) — Seven sections were found cracked on January 9th, at the cigar factory of Otto Eisenlohr and Brothers, Liberty and Ross Streets, Lancaster, Pa.

(34.) — Five sections of a heating boiler cracked January 10th, in the building at 1007-11 Market St., Philadelphia, Pa., owned by the estate of John Dobson.

(35.) — A tube ruptured January 10th at the National Metal Moulding Company's plant, Economy, Penn.

(36.) — A section ruptured January 11th, at the Children's Cottage of the Michigan State Sanitarium, Howell, Mich.

(37.) — Two sections of a heating boiler were found cracked January 12th, at the Franklin School, Creston, Ohio.

(38.) — A tube burst January 12th, at the cotton mill of the Arkwright Mills, Spartansburg, S. C.

(39.) — Two sections cracked January 12th, at the plant of W. D. Allen and Company, 5630-46 West 12th St., Chicago, Ill.

(40.) — A firebox section of a heating boiler cracked January 13th, at the garage of the U. S. Motor Sales Company, 621 First St., South Boston, Mass.

(41.) — A section of a heating boiler cracked on January 13th, in the apartment house of Peter and Andrew Bertelson, 1480 Larkin St., San Francisco, Calif.

(42.) — A section of a boiler cracked January 14th, at the bakery of the Consumers Baking Company, 76 Ferry St., Springfield, Mass.

(43.) — A section of a boiler cracked at the Old Burritt Public School, New Britain, Conn., on January 15th.

(44.) — On the 15th of January a hot water heating boiler exploded at the residence of Samuel Brownlie, near Hinsdale, Ill. The house was badly damaged, the property loss amounting to about $1,500.

(45.) — On January 16th a boiler exploded in a cooper mill at Fish, Mo., causing the death of Daniel E. Gunn.

(46.) — A heating boiler exploded on January 16th, at the residence of V. B. Huff, 72 North Main St., Geneva, N. Y.

(47.) — On January 16th, a locomotive exploded on the Galesburg and Ottumwa Division of the C., B. & Q. R. R. One man was killed.

(48.) — Two sections of a heating boiler were found to be cracked at the Franklin School, Creston, Iowa, on January 16th.

(49.) — On January 17th, a locomotive exploded near Horseheads, New York. The fireman was killed and the engineer was injured.

(50.) — A tube failed at the Gorge Power Plant of the Northern Ohio Traction and Light Company, Akron, Ohio, on January 17th.

(51.) — Two sections of a heating boiler cracked January 17th, at the main plant of the Western Printing and Lithographing Company, Racine, Wis.

(52.) — One tube was ruptured and two headers were cracked in an accident on January 17th, at the Nekoosa Edwards Paper Company, Nekoosa, Wis.

(53.) — One section of a heating boiler cracked January 17th, in a building at Broadway and Ferry Streets, Everett, Mass., owned by Sara A. Green.

(54.) — A tube in the power plant of the United Public Service Company, Rochester, Ind., ruptured on January 17th. One man was injured.

(55.) — Five sections of a heating boiler cracked January 18th, at the Buckland Building, Main Street, Woonsocket, R. I.

(56.) — A boiler in a greenhouse at Willow Grove, Pa., exploded on January 19th, and caused severe injury to one man.

(57.) — On January 19th, a locomotive exploded near Danville, Iowa, Three men were killed.

(58.) — On January 19th, a boiler exploded at the plant of the Interstate Iron and Steel Company at East Chicago, Indiana. Three men were killed and several others injured.

(59.) — Two men were injured when a fitting on a blowoff pipe failed January 9th, at the hat factory of the Hoyt Messinger Corporation, Danbury, Conn.

(60.) — A section cracked January 19th, at the property at 128 Southbridge St., Worcester, Mass., belonging to the R. C. Taylor Estate.

(61.) — A heating boiler exploded January 20th, in the garage of the Jewel Tea Company, Syracuse, N. Y.

(62.) — One man was injured when a tube failed January 20th, at the plant of the Nestle Food Company, South Dayton, N. Y.

(63.) — A blowoff pipe failed January 20th, in the office building of the Davidson Realty Company, 6th and Pierce Streets, Sioux City, Iowa.

(64.) — Two sections ruptured January 20th, in the office of the American Telephone and Telegraph Company, 21 Fifth St., Waterloo, Iowa.

(65.) — Two sections of a heating boiler cracked January 20th, at the apartment house of T. F. and A. K. Morrissey, 55 Imlay St., Hartford, Conn.

(66.) — On January 21st, a boiler exploded at the Alliance Dyeing Works, 80 East 13th Street, Paterson, N. J.

(67.) — One man was injured when a gauge glass broke January 21st, at the Fisk Street Station of the Commonwealth Edison Company, Chicago, Ill.

(68.) — Twenty-two sections were cracked at the garage of the Elton Motor Company, Youngstown, Ohio, on January 21st.

(69.) — A rupture occurred in the boiler at the plant of the Reliable Laundry & Dry Cleaning Company, 10th and Monroe Sts., Toledo, Ohio, on January 21st.

(70.) — A fire sheet ruptured in the power plant of the Home Lawn Mineral Spring Company, Martinsville, Ind., on January 21st.

(71.) — Five sections cracked January 23rd, at the Buckland Building, Main St., Woonsocket, R. I. This is a separate and distinct accident from No. 55 on January 18.

(72.) — Two sections were ruptured on January 24th, in the Public Library at Alliance, Neb.

(73.) — Nine headers cracked in the boiler room of the Philip Carey Mfg. Company, Lockland, Ohio, on January 24th.

(74.) — A tube ruptured in the hospital of the Sisters of Charity, Clifton and Dixmyth St., Cincinnati, Ohio, on January 24th.

(75.) — A water heater exploded, January 26th, at the home of M. L. Levitt, 1303 Northeast Boulevard, Philadelphia, Pa.

(76.) — On January 26th, a dye vat exploded at the plant of the General Processing Company, Collins & Willard Sts., Philadelphia, Pa. Two persons were injured.

(77.) — A section of a heating boiler cracked in the basement of the Manhattan Market Company, 594-600 Mass. Ave., Cambridge, Mass., on January 26th.

(78.) — One man was injured when a tube failed, January 26th, at the Alfred University, Alfred, N. Y.

(79.) — A tube burst January 26th, in the Augustine Mill of the Jessup and Moore Paper Company, Wilmington, Del.

(80.) — A fire sheet ruptured in the Adams Laundry, 1813 California Street, Omaha, Neb., on January 27th.

(81.) — Three sections were ruptured on January 28th, in the High School at Florence, Wis.

(82.) — On January 30th, a boiler exploded at Love's shingle mill, twelve miles south of Graceville, Fla. One man was killed and one injured.

(83.) — One man was killed when a tube burst, January 30th, at the plant of the New Albany Compress Company, New Albany, Miss.

(84.) — A tube burst at the power plant of the American Gas and Electric Company, Atlantic City, N. J., on January 31st.

(85.) — Six headers were cracked at the plant of the J. C. Ayer Company, Market Street, Lowell, Mass., on January 31st.

(86.) — On January 31st, a boiler exploded at the plant of the General Petroleum Company, Olinda, Calif. One man was killed.

FEBRUARY, 1920.

(87.) — A fire sheet bulged and ruptured at Fiske University, Nashville, Tenn., on February 1st.

(88.) — In an accident to a boiler at the Alfred University, Alfred, N. Y., on February 1st, one man was badly injured.

(89.) — On February 2nd, the water back in a kitchen range exploded at the home of Raymond Rich, Chester, Pa. Four persons were injured, one, a two year old child, seriously.

(90.) — One man was killed when a blowoff pipe failed, February 2nd, at the boiler house of the National Stove Co., Lorain, Ohio.

(91.) — Two sections of a heating boiler cracked, February 2nd, at the plant of the Fleischman Malting Co., Buffalo, N. Y.

(92.) — A section of a heating boiler belonging to Ralph Shulman, 312 South Warren St., Syracuse, N. Y., cracked on February 3rd.

(93.) — A fire sheet ruptured at the plant of the F. J. Lewis Mfg. Co., 2505 South Robey St., Chicago, Ill., on February 2nd.

(94.) — A section of a heating boiler cracked, February 3rd, at the office building of Jacob Horowitz, No. 9 Trumbull St., Worcester, Mass.

(95.) — Four sections of a heating boiler cracked, February 3rd, at the Rialto Theater, 404 E. Main St., East Rochester, N. Y.

(96.) — A tube failed at the Fritz Bros. Garage, Ripley. Ohio, on February 3rd.

(97.) — A fire sheet bulged and ruptured on February 3rd, at the power plant of the Williamstown Illuminating Co., Williamstown, Mich.

(98.) — On February 4th, a boiler exploded at the plant of the Albany Tanking Co., Albany, Ind. One man was injured.

(99.) — Three headers cracked, February 6th, at the plant of the American Steel & Wire Co., Waukegan, Ill.

(100.) — On February 6th, a boiler exploded at the plant of the Hayes Ionia Co., Grand Rapids, Mich. One man was killed and two others injured.

(101.) — On February 6th, a heating boiler exploded at the residence of E. McCormic, Ashley & Sigourney Sts., Hartford, Conn. The boiler was completely demolished and considerable damage done to the property.

(102.) — On February 7th, the boiler of a locomotive on the Western Maryland R. R. exploded at Clear Spring, Md. Three men were killed and three others injured.

(103.) — On February 8th a heating boiler exploded at the home of James Heinlein, Beach Cliff Road, Coraopolis Heights, Penna. Mrs. Heinlein was injured by a flying fragment of the boiler and died a few hours later.

(104.) — On February 8th, a compressed air tank exploded at the Midvale Steel Works, Philadelphia, Pa. One man was fatally injured.

(105.) — On February 9th, a boiler exploded at the Broad River sawmill near Black Mountain, N. C. Two men were killed and two others injured.

(106.) — On February 10th a vulcanizer exploded at the Five Points Automobile Co's garage, Durham, N. C. The explosion shattered the plate glass window in the front of the building and severely injured a school girl who was passing at the time. A man who was nearby also was injured.

(107.) — On February 10th, a boiler exploded at a sawmill near Chavies, Alabama. Two men were instantly killed. The mill was wrecked by the force of the explosion.

(108.) — Four sections of a heating boiler cracked, February 10th, at the nurseries of the Kemble Floral Co., Chariton, Iowa. The accident necessitated the installation of a new boiler.

(109.) — Four sections of a heating boiler belonging to the World Realty Co., 15th & Douglas Sts., Omaha, Neb., ruptured on February 11th.

(110.) — A valve body burst at the Nicetown (Phila.) Works of the Midvale Steel & Ordnance Works on February 11th. One man was killed.

(111.) — A tube ruptured, February 12th, at the plant of the Nichols Copper Co., Laurel Hill, Newton, Long Island, N. Y.

(112.) — A tube burst at the flour mill of the Schultz-Baujan Co., Beardstown, Ill., on February 12th. One man was injured and the property damage amounted to over $2,500.

(113.) — A section of a heating boiler belonging to L. Sinsheimer, 714 Broadway, New York City, cracked on February 13th.

(114.) — On February 13th, a boiler exploded at the plant of the Mason City Brick & Tile Co., Mason City, Iowa. One man was fatally injured.

(115.) — On February 13th, a locomotive exploded at Bend, Oregon, injuring four men. The locomotive was the property of the Brooks Scanlon Lumber Co. Damages exceeded $5,000.

The Hartford Steam Boiler Inspection and Insurance Company

ABSTRACT OF STATEMENT, JANUARY 1, 1920.
Capital Stock, . . . $2,000,000.00.

ASSETS.

Cash in offices and banks	$390,221.07
Real Estate	90,000.00
Mortgage and collateral loans	1,426,050.00
Bonds and stocks	5,702,983.62
Premiums in course of collection	597,171.35
Interest accrued	107,590.44
Total assets	**$8,314,216.48**

LIABILITIES.

Reserve for unearned premiums		$3,715,903.48
Reserve for losses		175,539.16
Reserve for taxes and other contingencies		401,420.50
Capital stock	$2,000,000.00	
Surplus over all liabilities	2,021,353.34	

Surplus to Policy-holders . .	**$4,021,353 34**
Total liabilities	$8,314,216.48 .

CHARLES S. BLAKE, President.

FRANCIS B. ALLEN, Vice-President,   W. R. C. CORSON, Secretary.

L. F. MIDDLEBROOK, Assistant Secretary.

E. SIDNEY BERRY, Assistant Secretary.

S. F. JETER, Chief Engineer.

H. E. HART, Supt. Engineering Dept.

F. M. FITCH, Auditor.

J. J. GRAHAM, Supt. of Agencies.

BOARD OF DIRECTORS

ATWOOD COLLINS, President, Security Trust Co., Hartford, Conn.

LUCIUS F. ROBINSON, Attorney, Hartford, Conn.

JOHN O. ENDERS, President, United States Bank, Hartford, Conn.

MORGAN B. BRAINARD, Vice-Pres. and Treasurer, Ætna Life Insurance Co., Hartford, Conn.

FRANCIS B. ALLEN, Vice-Pres., The Hartford Steam Boiler Inspection and Insurance Company.

CHARLES P. COOLEY, Hartford, Conn.

FRANCIS T. MAXWELL, President, The Hockanum Mills Company, Rockville, Conn.

HORACE B. CHENEY, Cheney Brothers Silk Manufacturers, South Manchester, Conn.

D. NEWTON BARNEY, Treasurer, The Hartford Electric Light Co., Hartford, Conn.

DR. GEORGE C. F. WILLIAMS, President and Treasurer, The Capewell Horse Nail Co., Hartford, Conn.

JOSEPH R. ENSIGN, President, The Ensign-Bickford Co., Simsbury, Conn.

EDWARD MILLIGAN, President, The Phœnix Insurance Co., Hartford, Conn.

EDWARD B. HATCH, President, The Johns-Pratt Co., Hartford, Conn.

MORGAN G. BULKELEY, JR., Ass't Treas., Ætna Life Ins. Co., Hartford, Conn.

CHARLES S. BLAKE, President, The Hartford Steam Boiler Inspection and Insurance Co.

The Locomotive

DEVOTED TO POWER PLANT PROTECTION

PUBLISHED QUARTERLY

VOL. XXXIII. HARTFORD, CONN., JANUARY, 1921. No. 5.

FLYWHEEL EXPLOSION AT STUTTGART, ARKANSAS.

Two Recent Flywheel Accidents.

THO we give it little thought, the old, old story, " — a kingdom was lost, all for the want of a horseshoe nail," is illustrated time and again in our daily life. Rather striking demonstrations of the time worn allegory were given recently by two flywheel accidents — the one at Stuttgart, Arkansas, which is illustrated on the front cover of this issue, and the other at Lancaster, South Carolina, from which the view above is taken.

The accident at the Stuttgart plant occurred on November 8th, 1920. The engineer, who was standing near the boiler room door, said he thought he noticed the engine speeding up and he started for the throttle to close it but before he reached the valve the flywheel burst. The flying parts of the rim and spokes went through the roof, which was completely demolished, and most of them landed on the ground outside. One large piece of the wheel, however, dropped through the roof of the pump room, broke a large gear wheel on a feed pump, and ruptured several large pipes so that it was necessary to shut off the city water supply. This left the plant without any water for the boilers and it was nearly two days before the damage to the water line was repaired and the boilers were put back into service. A large amount of shafting and electric wiring was also damaged.

On November 25th, an engine at Lancaster, South Carolina, started to overspeed while no one was in the engine room. Two men made an effort to reach the throttle and stop the engine before any damage was done but they were not quick enough and the flywheel burst. Part of the wheel ripped through the **roof of the** plant and seriously damaged the roof and porch of a house nearby. Other sections of **the** wheel were thrown in various directions damaging much of the plant equipment.

Both of these engines **were** of the Corliss type **and overspeed-ing resulted from breakage of parts** of the valve gear. To clearly describe the causes of the accidents reference will be had to some **sketches of the typical Corliss valve gear.** Fig. 1 shows the

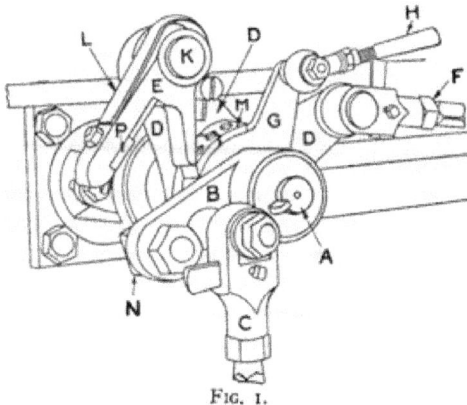

Fig. 1.

parts of this gear that are intimately related to and operate the steam admission valve for the head end of the cylinder. A similar arrangement is used at the crank end of the cylinder. In this sketch A is the end of the **valve-stem,** which passes through the other parts of the gear and into the cylinder where it is attached to the cylindrical Corliss valve. The valve is actuated by the oscillation of this **valve-stem.** Part B is the **steam-arm** and, as will be seen, is secured to the valve-stem by a key. The rod C, attached to the steam-arm, is drawn down to the position shown in the sketch by a **dash-pot** on its lower end which is not illustrated here. In the position shown, **the steam-valve is closed.** The **double-arm** D is machined to a running fit on the **valve-stem** and has two arms — one extending to **the** left and carrying the steam-**hook E and the** other extending to the right and having attached to it the **steam-valve-rod** F. This steam-valve-rod is attached, **through a series of levers,** to the eccentric of the engine and is **given a reciprocating motion which changes to an** oscillating motion in the **double-arm**. The **knock-off-lever** G is also machined to a running fit on the **valve-stem** and is controlled by the **governor-rod** H. The governor shifts the position of this knock-off-lever by rotating it about the center of the valve-stem but for any given load the position of this lever will be stationary.

The **steam-hook** E is carried by the **double-arm** to which it is fastened by the **pin** K although motion is permitted between the pin and the hook. A **spring** L is fastened at one end to the **double-arm** and at the other end exerts a pressure which tends to rotate the **steam-hook** in a counter clockwise direction. The **steam-hook** is prevented from turning, however, in the position shown, because the end of its short leg is bearing against the **knock-off-cam** M which is fastened to the **knock-off-lever** G.

Starting from the position shown the operation of the engine would be as follows: The **double-arm** will rock counter-clockwise so that the steam-hook will be moved downward. This will remove the steam-hook from the control of the **knock-off-cam** and the steam-hook will make a slight counter-clockwise rotation until stopped by the end of the short leg coming in contact with the circular part or body of the **knock-off-arm**. At the outer end and on the far side of the **valve-arm** will be seen the end of a rectangular block N which is secured to the arm. As the steam-hook swings down it will pass over this rectangular block and when the **steam-hook** has moved far enough the upper edge of the block P will snap under the lower edge of the block N. Soon after this occurs, the **double-arm** will begin to move in the opposite or clockwise direction and by the interlocking of the two blocks N and P the **steam-arm** will be lifted and the **valve-stem** and the **valve** will, in turn, be rotated to admit steam to the cylinder. This rotation of the **valve-stem** by the **steam-hook** and double-arm will continue until the point is reached where the short leg of the steam-hook strikes the knock-off-cam. This will rotate the steam-hook about its pivot K, the **valve-arm** will be released and the **dash-pot** will pull the valve shut. The governor, by altering the position of the **knock-off-arm**, will change the time of cut-off and will thereby control the speed of the engine.

It should be noted that in order to have the valve gear under the control of the governor the **steam-hook** must be tripped by the knock-off-cam before the **double-arm** begins its return or counter-clockwise motion. If tripping does not occur by that time it will not occur at all and the steam-valve will not close until late in the stroke of the engine — in fact, it will be very nearly full stroke before cut-off occurs. For the reasons given above it is necessary, in an engine with a single eccentric, that cut-off occur no later than about half stroke of the engine. The valve gear is therefore provided with an additional **safety-cam**, not visible in this sketch, on the **cam-arm**, that prevents the **steam-hook** picking up the valve-arm should the governor shift the **cam-arm** to the position of an extremely late cut-off. It should be noted that this safety-cam comes into play only when the governor is running slowly, either from breaking of the governor belt or an overload on the engine.

In the accident at Lancaster, the **knock-off-cam**, which is a hardened block secured to the **cam-arm**, usually by three screws, cracked through the middle screw hole as indicated in Fig. 1 and the lower half dropped off. The effect was as though no **knock-off-cam** were present and the load at the time was such that the engine overspeeded even though the governor shifted the **cam-lever** to an early cut-off position and in fact effected an early cut-off in the crank end of the cylinder.

In the accident at Stuttgart the **governor-rod** broke near the end throwing the **cam-arm** out of the control of the governor. The **cam-arm** then rotated with the **valve-stem** so that the **steam-hook**, not being tripped by the **knock-off-cam**, held on to the steam-arm and caused the engine to take steam for practically full stroke on the head end. The load was such that any governing of cut-off on the crank end of the engine was of no avail and the flywheel overspeeded and burst.

It would be difficult to say where the blame, if there is any, for these accidents should be placed. Mechanisms will wear out and breakage will occur, even in the best designs and when given the best of care. The safety of an engine depends upon the correct operation of a number of relatively small and often rather inaccessible parts — but even so, they are highly important parts. And the loss of a relatively unimportant horseshoe nail brought disaster to a horse, his rider and the kingdom.

Excessive Water as the Cause of Engine Breakage.

OF all the causes of accidental injury to steam engines, the presence of an extraordinary amount of water in the cylinders is one of the most prevalent. After the exhaust valve closes, this water, forced by the piston into the space remaining in the cylinder, has no ready means of escape. If the clearance space is quite large and the engine is running very slowly, there is a possibility of no damage resulting. If the water is sufficient to more than fill the clearance space, however, an irresistible pressure will be produced and something must give way to relieve it.

All engines are, or should be, provided with cylinder drains or drips, one at the throttle valve and one in each end of the cylinder, for the purpose of removing the water of condensation when starting. Ordinarily these drips are closed by means of a plain globe or angle valve, though in many cases the cylinder drips are fitted with spring loaded relief valves which may be opened also by hand. The principal safeguard that such spring loaded valves may afford is in the relieving of unduly high steam pressure. Contrary to a somewhat prevalent belief, valves of this type are of little or no use for draining the cylinder should water in any appreciable amount enter through either the steam or exhaust pipe while the engine is running. Water, at rest or when set in motion, has considerable inertia. When the piston drives a volume of water before it in its stroke, the water is flowing in lines parallel to the axis of the cylinder and a considerable force is required to turn it from this direction of flow. To expect it to make a sudden, right-angled turn and to converge all its lines of flow into a relatively small drain pipe is not reasonable. Even a relief valve in the center of the cylinder head would be wholly inadequate in a case of this kind.

Probably the most frequent result of excessive water in the engine cylinder is that the cylinder head is blown off. The head itself may be broken, the thread on the cylinder head studs may strip, or the studs or bolts themselves may pull apart, and not unfrequently the wreck may involve failure of all of these parts. Sometimes the cylinder itself may be cracked or badly broken and there are numerous instances of the excessive pressure having driven the piston on or off the piston rod. The cylinder and piston construction may be strong enough to withstand the shock and to transmit the force developed to other parts of the engine so that a piston rod may be broken or bent, a crosshead may fail, the connecting rod may be damaged, or the force may be transmitted still further to the crank and shaft of the engine so that these parts may suffer serious injury. Sometimes the shock is so great that the engine is a complete wreck. Water in an engine cylinder is, in fact, so serious a matter that every precaution should be taken to remove the possibility of its presence.

When the boiler feed water contains foreign substances in sufficient amount to cause excessive foaming and priming, there is great danger of water being carried over to the engines in amounts sufficiently large to cause damage. If the engine is some distance from the boiler and the steam pipe is exposed, there is also the possibility of excessive condensation taking place so that the engine may sometime receive a "dose" of water sufficiently large to wreck it. It is best, under any condition, to provide a separator in the steam line, which should be connected close to the engine throttle, not only for the purpose of removing excessive amounts of water but also for the extraction of small amounts of moisture in the steam. It is always in the interest of economy to supply the engine with as dry steam as possible.

The function of the steam separator is to separate the liquid from the vapor in the steam pipe but after this has been accomplished some means must be provided to drain the separator of the water. To do this a drain pipe is provided to conduct the water to a steam trap which will discharge the water but not the steam to the atmosphere. It is best to have as short and as direct a drain pipe as possible from the separator to the trap and numerous fittings and valves are to be avoided. If a valve is installed in the drain between the separator and the trap, it should be locked or sealed open. Steam traps, at times, will become inoperative from some cause and if such a condition prevails the separator, of course, is deprived of a drain so that it will accumulate water and will thereby become valueless and even dangerous. The steam trap, therefore, should be carefully watched and inspected to make sure that it is, at all times, in perfect working order.

There have been many cases of engine wreckage from water taken into the cylinder from the exhaust piping. When a cold engine is started the throttle valve is opened, of course, only a slight amount and there is the possibility of a vacuum forming by rapid condensation of the steam in the engine cylinder so that if, by chance, any water be present in the exhaust line there is great likelihood of its being drawn into the engine cylinder. Then, when the engine turns over, the water is trapped in the cylinder and is the cause of damage. To guard against the possibility of wreckage from the cause just mentioned precautions should be taken to prevent the accumulation of water in the exhaust pipe. From the standpoint of ease of removal of the water the best way would be to have the exhaust pipe so installed that it will act as its own drain, the water running by gravity to the open end of the pipe. In the great majority of cases, however, the exhaust pipe, before reaching an open end, rises above the engine cylinder so that a pocket is formed in which water may very readily accumulate. Such lines should be equipped with ample size drain pipes located at the lowest points of the exhaust pipe and the installation of these lines should be carefully studied to see that they are drains in fact and not additional pockets in the system.

When a cold engine is to be put into operation it is necessary that certain precautions be observed before it is set in motion. If the steam be admitted to the engine too rapidly, severe expansion stresses will be set up in the engine parts which might lead to cracking of the cylinders. Furthermore, when steam is first admitted to a cold engine, and until the machine becomes warmed, rapid condensation will take place with consequent formation of a sufficient amount of water to give rise, should the engine turn over before this condensation ceases or the water is removed, to a greater or less degree of shock. It is therefore quite essential that any steam engine, before it is started, be brought up to temperature slowly and gradually. The length of time required for this is naturally a factor depending entirely on the size of the engine.

With Corliss engines, or with any other type in which the valve gear can be operated by hand, the cylinder is readily warmed by "cracking" the throttle a little and turning, alternately, the steam valves at the head and at the crank end to the open position. The valves of some engines cannot be operated independently by hand and with these it is necessary to bar the engine over from time to time during the process of warming so that both ends of the cylinder may be heated at as nearly a uniform rate as is possible. It is hardly necessary to call attention to the necessity of closing the throttle before attempting to turn the engine over. These methods of warming the engine cylinder preparatory to starting are, of course, the necessary steps for a simple engine or the high pressure cylinder of a multiple expansion engine. In the case of a compound engine it is well to warm the receiver and low pressure cylinder by means of the high pressure steam bypass to these parts, when such bypass exists, or by turning the engine over several times by hand to permit the steam from the high pressure cylinder to flow on through the engine. By thus thoroughly warming the engine before starting the danger of producing a vacuum in the engine cylinder and thereby drawing any water from the exhaust pipe into the cylinder will be greatly reduced.

When shutting down an engine, there is a condition produced which makes the entrance of water into the engine cylinder from the exhaust pipe, should it be present there at such time in sufficient quantity, an almost absolute certainty. In order to understand how this condition is produced, reference will be had to some indicator diagrams. Fig. 1 shows a typical card or diagram for an average engine under load. The steam valve opens

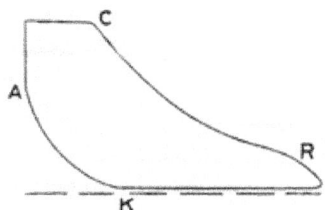

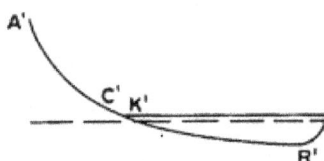

FIG. 1.   FIG. 2.

for the admission of steam at A and closes or cuts-off at C. The exhaust valve opens or releases at R and closes for compression of the steam remaining in the cylinder at K. Suppose now that the throttle of the engine is closed to stop the engine. On the next revolution, when the piston arrives at the point in its stroke indicated by K, there still will be some steam left in the cylinder and this steam will be compressed along the line K' A' of Fig. 2. When the piston starts in the opposite direction this steam will re-expand. At the point A' the steam valve opens and the indicator diagram is therefore dealing with a greater volume of steam (the volume in the cylinder plus the volume in the steam chest made available by the opening of the admission valve) so that it is difficult to say just what line this re-expansion will follow. However, since the volume is increasing by the forward motion of the piston and since the steam is condensing somewhat, it is certain that the pressure will fall and it will be fairly accurate for the purpose of this discussion to say that it follows back along the line A' K'. After the piston reaches the point in its stroke where the steam valve closes, or about at C', the steam in the cylinder will continue to expand so that the pressure will drop considerably below atmospheric. We thus have a rather low cylinder pressure when the piston reaches the point where the exhaust valve opens. If a considerable quantity of water has accumulated in the exhaust pipe from any such cause as, for instance, the premature stopping of the pump of a jet condenser, this water may readily be drawn up with a rush into the engine cylinder. This rising of water into the cylinder in an attempt to equalize the cylinder pressure with that of the atmosphere results in the upward curve of the indicator card at the point R'.

When the engine makes the return stroke with the exhaust valve open, the water is carried ahead of the piston into the end of the cylinder. Some of it, of course, will return to the exhaust pipe, but some of it is sure to remain in the cylinder. When the water was drawn into the cylinder there were forces acting in a very positive direction to move it. After it enters the cylinder these forces are not so definite in their direction and the inevitable result will be that considerable water will be trapped in the cylinder after the exhaust-valve closes. With the engine still turning over at close to its normal speed and with no escape for the water, the result is obvious.

With condenser operation, it is rather common practice to install an automatic device on the condenser for the purpose of "breaking" the vacuum should the condensing water rise into the exhaust pipe. These vacuum breakers consist, quite frequently, of a float inside the condenser which, when the water rises, opens a valve to admit air to the condenser above the water level so that the further influx of injection water is prevented. This principle of breaking the vacuum is entirely correct but accidents have occurred from the float becoming water logged or from sticking of the valve. If the air is not admitted in sufficient quantity to break the vacuum before the water reaches the level where the air is admitted, the danger is accentuated, for if the water passes the point of air admission the air, when it does enter, will mix with the rising column of water, which will thereby have its specific volume greatly decreased, and it will be possible for the mixture to rise to a higher level than could the liquid alone. In order to avoid the condition just mentioned, it is well to admit the air as close as possible to the engine itself and to use a vacuum breaker that is reasonably positive in its action.

A condition that is sometimes encountered and one that is extremely dangerous is that in which the injection water to the condenser is supplied under a pressure head. With this condition it is possible for the water to flow into the engine cylinder while the engine is at rest so that the machine may be wrecked upon starting up. The water should always be drawn up to the condenser by the vacuum or else the head under which it flows to the condenser should be so small that, with the vacuum removed, the water would not rise to a point dangerously near the engine.

There may be cases in which the water is supplied the condenser from a pond, the level of which at the time of observation or inspection may not be dangerous but which may at times be subject to a large fluctuation from causes such as a heavy rainfall or rising tides and an unsafe condition may thereby be produced.

A very good preventive against the entry of water from the exhaust side of the engine is some form of non-return valve which will close when the water flows toward the cylinder. Such an installation is illustrated on page 140. It is always best, of course, to have an atmospheric relief valve in the exhaust line to prevent excessive pressures building up in the low pressure cylinder of a multiple expansion engine and also, of course, to prevent immediate shut-down should the condenser equipment cease to operate. The connection for this relief valve should be between the engine and the non-return valve as indicated in the illustration.

Not only jet but also surface condensers may give trouble if the latter type is so connected that the steam enters the condenser shell at the bottom. This method of piping is sometimes used when there is no basement under the engine and the condenser is placed on a level with the engine cylinder. If the condensate pump stops accidentally, the exhaust pipe will accumulate water from the condensing steam and at the time that the engine is being shut down this water may return to the cylinder to cause damage as explained in the foregoing.

Still another source from which water may enter an engine cylinder is the receiver on a compound engine. Engines of this type are sometimes constructed with a valve gear giving a fixed

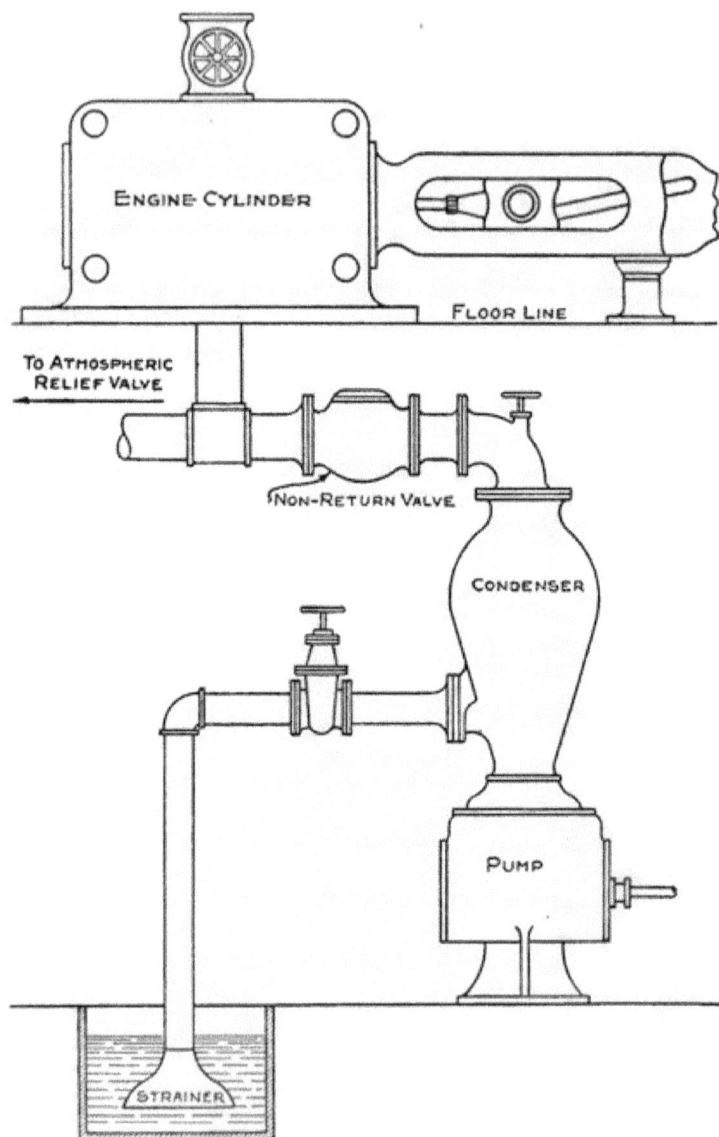

ENGINE CYLINDER

FLOOR LINE

TO ATMOSPHERIC
RELIEF VALVE

NON-RETURN VALVE

CONDENSER

PUMP

STRAINER

TYPICAL JET-CONDENSER INSTALLATION.

cut-off to the low-pressure cylinder and are governed by a variable cut-off in the high-pressure cylinder. With a very early cut-off in the high-pressure cylinder the low-pressure cylinder acts as a pump, drawing the steam from the receiver, and this action may continue to the point where the pressure in the receiver will be lower than atmospheric. This is particularly true of an engine operating condensing. The receivers of compound engines are, of course, or at least should be, fitted with drain traps but as these are sometimes operative only when there is a pressure in the receiver greater than atmospheric the condition outlined above will cause such a trap to cease its functioning. Inasmuch as continued operation under these conditions must result in the accumulation of condensation in the receiver, it is not long before the low pressure cylinder receives a harmful " dose " of water. Of course, the proper apparatus to use for the draining of engine receivers is a special form of trap which will operate under vacuum or pressure so that some means of removing the water in provided under all conditions.

From what has been said above it is evident that there are a number of ways and reasons for the entry of water into an engine cylinder. All of these hazards may be mitigated by the use of the proper apparatus but it must never be forgotten that such safeguards as are installed must be subjected to the most rigid inspection as to the construction and method of installation and that careful watching is needed at all times to obtain dependable service.

With Reference to the Changing of Old Lap Seam Boilers to a Butt Seam Construction.

NOT long after the steam boiler came into general use in America, considerable discussion was aroused with respect to the question of limiting the life of a boiler. Numerous instances of serious accident, which it seemed impossible to account for, had impressed many with the idea that a boiler, like any other piece of apparatus, was subject to deterioration from constant use and that therefore it would be best to take a boiler out of service after a certain period. In fact, a number of concerns followed this practice. The majority of boiler users and engineers, however, felt then as they do now that rigid inspection would safeguard their boiler plants and would furthermore be of greater service in the interest of economy, for it was admitted that many boilers had served for twice the life that, by some, had been allowed for safety.

The plan of relying on inspection for a forewarning was adopted and served well but there were a number of unaccountable explosions in boilers of relatively short life. At the time, the majority of boilers in use were constructed with a longitudinal lap joint. A series of investigations was conducted to study the stress conditions in this type of joint and it was found that the construction, both from its fundamental shape and the conditions of manufacture, presented a most dangerous condition.

In The Locomotive for April 1905 there appeared an account of the disastrous boiler explosion at Brockton, Mass., on March 20, 1905, and also an article on the "Lap-joint Crack" to which type of defect the explosion was said to be due. For the sake of clearness we shall present here some of the more important points which were brought out in the last named article.

When a boiler plate is rolled to a cylindrical form, the edges of the plate, in passing through the rolls, are not gripped as effectively as is the middle of the plate so that the ends are left somewhat flat. The condition produced is illustrated in Fig. 1. This necessitates the plates being forced together at the edges and this produces an added stress that persists unless relieved

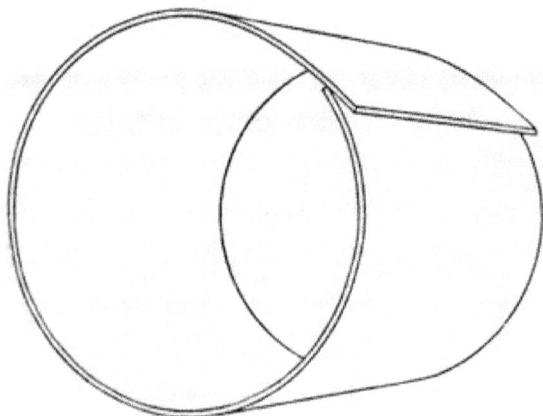

FIG. 1.

by annealing. In addition to this the plates, if bent after punching, will bend along a line of rivet holes as shown in Fig 2 in somewhat exaggerated form.

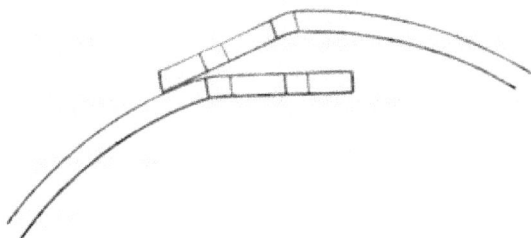

FIG. 2.

The elementary lap joint is illustrated in Fig. 3. If tension is applied as indicated in Fig. 4, the plates, in an attempt to align themselves with the load, will bend along a line running under the outer edge of the rivet heads.

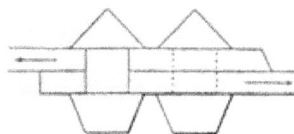 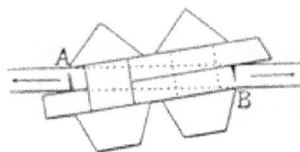

FIG. 3.   FIG. 4.

The combined effect of all these conditions, together with the constant bending of these joints by changes of pressure when in use, is to impose excessive stresses in the surface of the boiler plate along the line just mentioned. This has produced, in many boilers, a crack which starts always from the inside or covered surface of the inside or the outside plate of the joint as indicated at A and B in Fig. 4. This crack may eventually work its way through the plate until it shows itself by leakage. But in many cases it may develop for some distance along the joint and yet remain absolutely invisible. Eventually the weakness may develop to the point of complete failure and a disastrous explosion.

Inspection is generally accepted as being safe for the determination of the fitness of a boiler for use. The lap seam crack, however, is invisible to all methods of inspection, except cutting out the rivets and separating the plates or the method described in The Locomotive for October 1914. Recognizing the insidious danger presented in the lap joint for longitudinal seams in boilers, the Boiler Code Committee of the American Society of Mechanical Engineers formulated the following regulation: (*Par.* 380, *A.S.M.E. Boiler Code, Edition* 1918.)

"*The age limit of a horizontal return-tubular boiler having a longitudinal lap joint and carrying over 50 lbs. pressure shall be 20 years, except that no lap joint boiler shall be discontinued from service solely on account of age until 5 years after these rules become effective.*"

Some boiler owners may be of the opinion that the longitudinal lap seams of boilers of this type can be changed to butt strap construction and the boilers kept in service after the time limit. This change of design and construction is not approved, however, by those thoroughly familiar with steam boilers for, although butt straps and more rivets may be added, the material along the line of the joint, which was abused and tortured by the forming of the lap joint and fatigued by the years of service which subjected it to the expansion and contraction brought about by the many changes of temperature and pressure, would be further abused on the portion of the original construction left after cutting off one side of the lap joint and forcing the edges of the plate into line to form a butt joint.

Assuming that a double riveted lap joint has been changed to a triple riveted butt construction, the joint, after placing the butt straps and riveting, would appear as shown in Fig. 5 on page 145 with the rivet holes of the original lap joint at B and the new holes at C. If a defect existed in the plate as shown at A the joint would be very faulty. Assuming the original joint as having the rivet holes spaced 3¼ inches and the additional holes spaced 6½ inches, if the plate material were defective or contained a lap crack as shown, the failure of the joint would require only the shearing of the rivets in long pitch, or 6½ inches, and the failure of the defective plate.

It might be argued that the exposure of the inside of the lap seam, when the change to a butt seam is made, would reveal the presence of a lap seam crack. A crack of this nature is, however, often present in a boiler of this construction after years of service although it may not be visible to the naked eye. But, even though no crack exists,

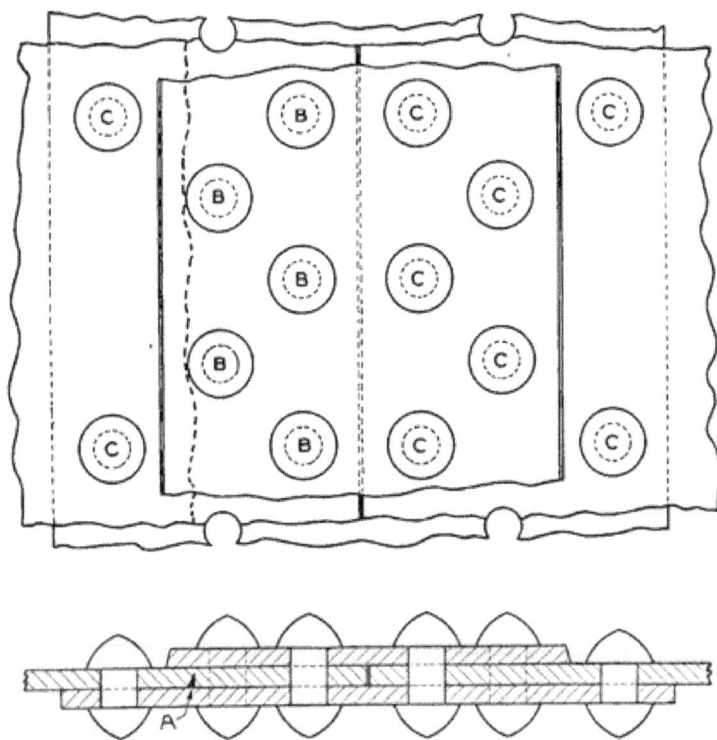

FIG. 5.

it must be remembered that boiler plate, like any other material, becomes fatigued after long years of service and for this reason, after it has been under stress for many years, it should not be subjected to a change of shape and torture of the material in an endeavor to keep the boiler in service, especially when there is evidence that the altered structure is defective.

The age limit of twenty years is none too exacting as will be evidenced by an explosion, resulting from a lap seam crack, of a boiler at the Tallahoma Lumber Co., at Mossville, Miss., on October 21st, 1920. This boiler was less than five years old. The explosion completely wrecked the plant, killed three men and injured four others. There was no negligence on the part of the operators and there was ample proof that the accident did not result from low water or overheating. On the other hand, the lap seam cracks could be clearly seen in the boiler plates after the disaster.

Regulations such as we have quoted are not intended to be arbitrary. Railroad companies determine the **safe** load capacity for each of their freight cars and if they discover an overloaded car they refuse to transport it. This is done not only to avoid the possibility of straining or breaking the overloaded car but also to prevent a possible wreck which might result in loss of life, property damage, and delay. In a like manner the Boiler Code Committee requests that steam boilers and pressure vessels be designed and constructed for a safe working pressure and that they be not subjected to overloads. The rule quoted above is sane and economic because it is intended for the protection of life, limb and property. So also should be regarded the action of the boiler inspector in condemning any construction regarded as unsafe.

Guarding the Coal Pile.

IT repeatedly has been said that no better opportunity exists for saving or for waste than in the coal pile. Modern improvements have brought the **engine and** the steam turbine to a high degree of perfection. If the exhaust steam is used to heat the feed water and for industrial purposes, there is practically no waste of the heat in the plant as a whole. But if we impose improper conditions on our boilers we might just as well store our coal on an open lot with a a sign reading, "Help Yourself."

Possibly both you and your chief engineer know what constitute proper conditions. But you are not operating the boiler plant. It is the fireman who holds the power to lock or unlock the back door to your coal pile. Experience is the usual school for firemen and while that school teaches its lessons well, when once they are grasped, its demonstrations are often, for the firemen, hidden and obscure. And, because of this, little of the fundamental facts is handed down from fireman to fireman and, in consequence, he does not progress in firing efficiency as rapidly as he might. And yet he is trying hard to grasp the facts and the reasons. It is not natural for the human mind, unless it be of the lowest order, to take no interest in its work, to solve no problems and to make no progress.

Is it not our duty, not only to ourselves but to our firemen, to see that they are given the proper knowledge to help them to fire efficiently? The Hartford Correspondence Course for Firemen was developed for just this purpose. Written in a clear and interesting style, it traces and solves the problem of combustion from the simplest illustrations to the more obscure facts. The burning of volatile matter and the change of carbon to carbon dioxide or to carbon monoxide are discussed in a manner easily understood and the methods for obtaining complete combustion are clearly explained. The use of the draft gauge and the care of the fire are given full treatment. Seventeen lessons of the course are devoted to this very important problem of efficient combustion. Seven lessons follow this, dealing with the proper care of the boiler to insure its being in condition to absorb the heat of the furnace and the precautions to be taken for its safety.

The lessons are issued one at a time. When a student has gone over a lesson and sent in his answers to the questions on that lesson the next installment is sent him, together with a friendly, helpful letter of encouragement and advice. When the twenty-four lessons of the course have been completed, a final examination is given and when that has been satisfactorily passed the student is given a certificate of completion of the work.

We believe it an advisable plan to have the student pay the expense of the course himself. The charge is small, being but five dollars. It has been the practice of many employers, when the student has secured his certificate, to refund the man his money as evidence of their interest in his endeavors. The principle involved is that the student, having made an outlay of his own money, has his interest the more thoroughly aroused and is the more eager to complete the work.

We have several hundred men enrolled in the course and find them to be eager for knowledge and very enthusiastic in their praise of the work. These men range from Superintendents and Chief Engineers down to the most inexperienced firemen.

Enrollment blanks for the course may be had on application to any of our offices and further information will be cheerfully given upon writing to the Hartford Office of the Company, attention of the Correspondence Course Department.

The Locomotive

DEVOTED TO POWER PLANT PROTECTION

PUBLISHED QUARTERLY

WM. D. HALSEY, EDITOR.

HARTFORD, JANUARY, 1921.

SINGLE COPIES *can be obtained free by calling at any of the company's agencies.*
Subscription price 50 cents per year when mailed from this office.
Recent bound volumes one dollar each. Earlier ones two dollars.
Reprinting matter from this paper is permitted if credited to
THE LOCOMOTIVE OF THE HARTFORD STEAM BOILER I. & I. CO.

THE reports of boiler explosions that appeared in the early issues of The Locomotive were comparatively lengthy articles and not infrequently extended through a dozen or more lines of type. At that time boiler explosions were far less numerous than at present so that a detailed description could be given of each one without encroaching too much on the other reading matter. As time went on, however, and the number of explosions increased, it became necessary to shorten the articles very materially and in the past few years little other than the date, nature of accident, owner, and location has been given. Even this cutting of detail, however, has been inadequate, of late, to reduce the space required to list all the accidents. To decrease the space required it has become necessary to go still further, and in this issue, as will be noted by reference to page 151, we are adopting a tabular method of presentation. It is believed that this method will give all the information of interest and will, at the same time, make these lists better adapted to a search for data under any particular heading.

President Blake Celebrates Sixtieth Birthday.

WHEN President Blake entered his office on October 25th, a vase containing sixty roses, and a stack of telegrams and letters reminded him that another birthday, his sixtieth, had rolled around.

Later in the day the Home Office force gathered in President Blake's office to express to him their best wishes for the day. At this time the officers, managers, and general agents of the company presented Mr. Blake with a silver service and announcement was made that the field force had made the day the culmination of a campaign for the greatest week of business ever done by the company, amounting to more than six times the weekly average for the past year.

In the evening a surprise reception in Mr. Blake's honor was given at his home and the employees of the company took advantage of the opportunity to offer him their felicitations and best wishes for many more years of a full and useful life.

Personal.

The Hartford Steam Boiler Inspection & Insurance Company naturally is happy to publish to its friends any fact or information which redounds to its credit, even that which touches it only as glory reflected from the achievements of a member of its organization. We therefore grasp this, our first opportunity to record with pride the distinguished honor that has been conferred on our resident agent in Minnesota, Mr. T. W. Hugo of Duluth During the past summer the mayor of that city resigned his office to accept the nomination of Judge of the District Court. It then, it appears, was incumbent on the commissioners of Duluth to select a man to fill the vacancy. The Duluth Herald thus explains the result of their deliberations:—

> "The commissioners, in searching for a man who would meet the approval of the majority of the citizens, agree that T. W. Hugo would best fit the requirements. A number of other names were suggested but none received the approval of all the commissioners except Mr. Hugo."

We readily agree with the commissioners and commend their choice. We know Mr. Hugo from an association of over thirty-five years as a man of exceptional ability and tested integrity. Sincere and frank of character, he has given to our Company a loyal devotion which has not been displayed only by passive acquiescence in its policies and activities but has extended also to vigorous and constructive criticism at times when he felt its proposals were not to its best interest or those of its policyholders. We can readily appreciate, therefore, the kind of a citizen he has been been in Duluth and that the interest he has taken in public matters has been characterized by an aggressive activity for the good of the community that has won the respect and regard of his fellow townsmen. He has served his city as alderman, President of its Council, member and President of its Board of Education and Chairman of its Committee of Public Affairs. In 1900 he was elected mayor and administered the city government for a term of four years thereafter. That he was a good and faithful servant in all these capacities is now proved by this unanimous action of the commissioners. We congratulate the city of Duluth in the selection made and extend to the new mayor our felicitations and best wishes for a successful administration.

BOILER EXPLOSIONS.

(INCLUDING FRACTURES AND RUPTURES OF PRESSURE VESSELS)

MONTH OF FEBRUARY, 1920
Continued from October, 1920 Issue

No.	DAY	NATURE OF ACCIDENT	Killed	Injured	CONCERN	BUSINESS	LOCATION
116	14	Boiler exploded	3	6	Puget Sound Traction Light & Power Co.	Power Plant	Holyoke, Mass.
117	14	Top head of vertical boiler failed			Nansemond Brick Corp'n	Brick Plant	Norfolk, Va.
118	14	Stop valve ruptured			Houston Light & Power Co.	Power Plant	Houston, Texas.
119	15	Section of heating boiler cracked			Adolph Abbey	Apt. House	St. Louis, Mo.
120	15	Section of heating boiler cracked			J. F. Dolin	Residence	Hartford, Conn.
121	15	Section of heating boiler cracked			Staunton Ave. School	School	Piqua, Ohio.
122	16	Tube failed			Northern Ohio Traction & Lt. Co.	Power Plant	Cuyahoga Falls, O.
123	16	Boiler ruptured			Ohio River Sand Co.	Dredge	Louisville, Ky.
124	17	Three tubes pulled out			Grasselli Chemical Co.	Chemical Plant	Grasselli, Ind.
125	17	Stay brace failed			City of Lansing	Heat & Power	Lansing, Mich.
126	18	Boiler exploded			Michael Lambert	Residence	Philadelphia, Pa.
127	18	Tube ruptured			Consumers Power Co.	Power Plant	Flint, Mich.
128	18	Tube ruptured			Industrial Works	Industrial Mach.	Bay City, Mich.
129	18	Accident to a boiler			H. C. Frick Coke Co.	Coke Plant	Calumet, Pa.
130	19	Blow-off pipe fitting failed			Armour & Company	Meat Packers	New York, N. Y.
131	20	Section of heating boiler cracked			Louisville Industrial School	School	Louisville, Ky.
132	21	Boiler exploded			B. E. Sassman	Greenhouse	Elkhart, Ind.
133	21	Hot water heating boiler exploded			Sunset Building	Office Bldg.	Bellingham, Wash.
134	22	Accident to a boiler			Y. M. C. A.	Y. M. C. A.	Wichita, Kan.
135	22	Drum of boiler bulged, tubes warped and pulled out			Dodge Bros.	Mill Supplies	Hamtramck, Mich.
136	23	Flue blew out			American Coating Mills	Paper Mills	Elkhart, Ind.
137	23	Accident to a steam pipe	2		U. S. Government	Steamship "Kitty"	
138	23	Section of heating boiler cracked			Blakesville Forging Co.	Forgings	Plantsville, Conn.
139	24	Accident to a boiler			Charlestown Mining & Mfg. Co.	At Washer No. 2	Ft. Meade, Fla.
140	24	Header of boiler cracked			Lack. & Wyoming Valley R. R.	Power Plant	Scranton, Pa.
141	24	Section of heating boiler cracked			Henry Ford Public School	School	H'land Park, Mich.
142	26	Two sections heating boiler cracked			Adolph Abbey	Apt. House	St. Louis, Mo.

MONTH OF FEBRUARY, 1920 (Continued)

No.	DAY	NATURE OF ACCIDENT	Killed	Injured	CONCERN	BUSINESS	LOCATION
143	26	Section of heating boiler cracked			Harmony Cafeteria	Cafeteria	Omaha, Nebr.
144	26	Fire sheet of boiler cracked			Tekemah Electric Light Plant	Power Plant	Tekemah, Nebr.
145	27	Accident to a boiler			Columbia Chemical Co.	Chemical Plant	Barberton, Ohio.
146	27	Blow-off pipe failed			Alliance Cold Storage & Packing	Meat Packers	Alliance, Ohio.
147	27	Fire sheet bulged, crown sheet was distorted and large number of stays failed			Clinton Cotton Mills	Cotton Mill	Clinton, S. C.
148	28	Section of heating boiler cracked			Beaver Falls Public School	School	Beaver Falls, Pa.
149	28	Tube ruptured			Baton Rouge Elec. Co.	Power Plant	Baton Rouge, La.
150	29	Nine sections of heating boiler cracked			Auditorium Amusement Company	Theatre	Malden, Mass.
151	29	Tube and header failed			The Congoleum Co.	Floor coverings	Marcus Hook, Pa.
152	29	Tube sheets bulged and seams were sprung			Salmen Brick & Lumber Co.	Brick Plant	Slidell, La.

MONTH OF MARCH, 1920

No.	DAY	NATURE OF ACCIDENT	Killed	Injured	CONCERN	BUSINESS	LOCATION
153	1	Section of heating boiler cracked			Mass. Baking Co.	Bakery	Holyoke, Mass.
154	1	Four headers cracked			Acme Wet Wash Laundry	Laundry	Canton, Ohio.
155	2	Two sections heating boiler cracked			L. Sovrensky	Junk Dealer	E. C'bridge, Mass.
156	3	Two railroad locomotives exploded		3	C. R. R. of N. J.	Railroad	Elizabethport, N.J
157	3	Header pulled out of fitting			Iowa Southern Utilities Co.	Power Station	Centerville, Iowa.
158	3	Four headers cracked			Amer. Steel & Wire Co.	Steel Plant	Waukegan, Ill.
159	4	Tube ruptured			Mac Sim Bar Paper Co.	Paper Mfgrs.	Otsego, Mich.
160	4	Pipe nipple pulled out of union			Amer. Distilling Co.	Distillery	Pekin, Ill.
161	4	Tube failed			Cleveland Builders Supply & Brick Co.	Brick Plant	Cleveland, Ohio.
162	5	Water arch ruptured			Baldwin Avenue Hospital	Hospital	Jersey City, N. J.
163	5	Section of heating boiler cracked			Vonhoff Hotel	Hotel	Mansfield, Ohio.
164	5	Section of heating boiler cracked			Jas. J. O'Meara	Residence	New York City.
165	5	Boiler exploded		12	Cross Keys Restaurant	Restaurant	Philadelphia, Pa.
166	5	Boiler sheet bulged			C. R. & J. B. Cline	Flour Mill	Cameron, Ohio.
167	6	Boiler of locomotive exploded		2	Birmingham Belt R. R.	Railroad	Birmingham, Ala.
168	6	Pipe fitting failed		1	N. Y. N. H. & H. R. R.	Railroad	Waterbury, Conn.

No.	Day	Nature of failure	Owner		Character of plant	Location
169	7	Tube ruptured			Lumber	Tacoma, Wash.
170	8	Tube ruptured	St. Paul & Tacoma Lumber Co.		Steel Plant	Lackawanna, N. Y.
171	8	Section heating boiler cracked	Lackawanna Steel Company	2	Office Building	Laurel, Miss.
172	9	Steam header cracked	City Hall		Mfg. of Mall. Cast.	Chicago, Ill.
173	9	Bank of tubes failed	National Malleable Castings Co.		Chemical Works	Barberton, Ohio
174	9	Tube ruptured	Columbia Chemical Co.		Steel Works	Lockport, N. Y.
175	10	Tube ruptured	Simonds Mfg. Co.		Brick Plant	Pittsburgh, Pa.
176	10	Two sections of heating boiler cracked	J. H. Ward & Sons Co.	1	Loft Building	New York City
177	10	Elbow in pipe line failed	Jacob Singer		Piano Case Mfgrs.	Arlington, Mass.
178	11	Two headers cracked	Theodore Schwamb Co.		Steel Plant	Waukegan, Ill.
179	12	Boiler bulged and cracked	Amer. Steel & Wire Co.		Flour Mill	Dalton, Mich.
180	14	Tube ruptured	Dalton Milling Co.		Axle Works	Canton, Ohio
181	15	Section of heating boiler cracked	Timken Detroit Axle Co.		Sanitarium	Columbus, Ohio.
182	16	Boiler bagged and ruptured	Columbus Sanitarium	1	Brick Plant	Nelsonville, Ohio.
183	16	Tube pulled out	Hocking Valley Brick Co.		Power Plant	Okmulgee, Okla.
184	16	Accident to blow-off pipe	Okmulgee Ice & Light Co.		Silk Mill	Harriell, N. Y.
185	16	Tube ruptured	Merrill Silk Company		Lumber	Eugene, Oregon
186	19	Four sections of heating boiler cracked	Booth, Kelly Lumber Co.		Factory	Brooklyn, N. Y.
187	19	Fire sheet bulged and ruptured	Robert Findlay Mfg. Co.		Dairy	Rockford, Ill.
188	19	Section of heating boiler cracked	Union Dairy Co.		Theatre	New York City
189	19	Fire sheet bulged and ruptured	Ansonia Amusement Co.		Ice Plant	Nashville, Tenn.
190	19	Boiler exploded	Atlantic Ice & Coal Co.		Hoisting Engine	Fairchance, Pa.
191	19	Three sections of heating boiler cracked	Unknown	1	Loft Building	New York City.
192	21	Crown sheet pulled from stays	N. & D. Kramic		Lumber	Foley, Ala.
193	23	Tube ruptured	Miller Brent Lumber Co.		Steel Plant	Waukegan, Ill.
194	23	Fine sheet of boiler cracked	Amer. Steel & Wire Co.		Factory	Moline, Ill.
195	23	Boiler exploded	Moline Rock Island Mfg. Co.		Clean. & Dye Wks.	Los Angeles, Cal.
196	24	Boiler sheet bulged and ruptured	Mission Cleaning & Dye Works		Ice Plant	Dallas, Texas.
197	26	Two sections of heating boiler cracked	Pine Ice & Cold Storage Co.		Paper Boxes	Providence, R. I.
198	26	Tube ruptured	Jencks Paper Box Co.		Paper Mill	Kalamazoo, Mich.
199	26	Tube ruptured	Bryant Paper Company		Rolling Mill	Fort Dodge, Iowa.
200	26	Four headers fractured	Fort Dodge Culvert & Iron Mills		Power Plant	Harrisburg, Ill.
201	27	Tube ruptured	Middle West Utilities Co.		Steel Plant	Waukegan, Ill.
202	28	Boiler of locomotive exploded	Amer. Steel & Wire Co.		Railroad	Chicago, Ill.
203	29	Three headers cracked	C. M. & St. P. R. R.	2	Chemical Works	St. Louis, Mo.
204	30	Section of heating boiler cracked	Monsanto Chemical Works		Apartment House	Louisville, Ky.
205	30	Pipe fitting on boiler failed	Mrs. L. V. Turner		Coal Dealers	Pittsburgh, Pa.
			Pittsburgh Coal Company			

MONTH OF APRIL, 1920

No.	DAY	NATURE OF ACCIDENT	Killed	Injured	CONCERN	BUSINESS	LOCATION
206	2	Section of heating boiler cracked			Charlesbank Homes, Inc.	Real Estate	Boston, Mass.
207	2	Tube ruptured			Swift & Company	Meat Packers	Chicago, Ill.
208	6	Tube ruptured			Columbia Chemical Co.	Chemical Plant	Barberton, Ohio.
209	6	Fire sheet bulged and cracked			Village of Arcade	Municipality	Arcade, N. Y.
210	6	Tube ruptured			Consumers Power Co.	Power Plant	Flint, Mich.
211	6	Section of heating boiler cracked		1	Lewis Brothers	Apt. House	Tacoma, Wash.
212	6	Failure of blow-off pipe			National Gas & Heating Co.	Power Plant	Nashville, Tenn.
213	6	Boiler exploded			Keystone Woodwk & Supply Co.	Mill Supplies	Philadelphia, Pa.
214	7	Tube burst			Philadelphia Elec. Co.	Power Plant	Philadelphia, Pa.
215	7	Boiler on hoisting engine exploded		3	Manhattan Construction Co.	Contractors	Okla. City, Okla.
216	8	Safety valve broke		1	J. Obenberger Forge Co.	Drop Forgings	West Allis, Wis.
217	9	Section of heating boiler cracked			Wendt & Rausch Company	Printers	Toledo, Ohio.
218	9	Two sections of heating boiler cracked			A. C. Johnson	Residence	Nevada, Mo.
219	11	Boiler exploded			D. R. Morrow	Drilling for Oil	Denver, Colorado.
220	12	Seven headers cracked in W. T. Boiler			Houston Gas & Fuel Co.	Gas Plant	Houston, Texas.
221	12	Pipe fitting failed		1	Cincinnati Frog & Switch Co.	R. R. Switches	Cincinnati, Ohio.
222	13	Boiler exploded			Harbison & Walker Refract. Co.	Fire Brick	Clearfield, Pa.
223	13	Boiler exploded			E. Reingnach	Bakery	Coalinga, Cal.
224	13	Section of heating boiler cracked			L. V. Niles	Apt. House	Boston, Mass.
225	13	Crown sheet bagged in locomotive			W. A. Bechtel	Contractor	San Fran., Cal.
226	14	Tube ruptured			Anderson & Middleton Lumber	Lumber	Aberdeen, Wash.
227	14	Boiler exploded		1		Apt. House	L. I. City, N. Y.
228	15	Three sections heating boiler cracked			Chevrolet Motor Co.	Auto Mfgrs.	New York, N. Y.
229	15	Sections of heating boiler cracked			Dr. M. Golland	Residence	St. Louis, Mo.
230	15	Two sections of heating boiler cracked			Board of Education	School	Travene City, Mo.
231	15	Boiler exploded			U. S. Shipbldg. Emer. Fleet Corp.	Shipbuilding	Mobile, Ala.
232	16	Two tubes in heating boiler cracked			Charlesbank Homes	Apt. House	Boston, Mass.
233	19	One tube burst, 95 tubes damaged			Michigan Alkali Co.	Chemical Works	Wyandotte, Mich.
234	20	Boiler exploded			L. Crossman	Oil Drilling	West Plains, Mo.
235	20	Boiler exploded, resulting fire did $20,-000 damage					
236	21	Shell plate damaged			A. Ribernio	Steam & Vul.	Riverhead, N. Y.
237	22	Section of heating boiler cracked			Cresent Laundry Co.	Laundry	Webb City, Mo.
238	23	Boiler of locomotive exploded		3	Borough of Indiana	School	Indiana, Pa.
					Booth-Kelley Lumber Co.	Lumber Camp	Eugene, Ore.

No.	Day	Nature of Accident	Owner	Establishment	Location
239	23	Tube ruptured	American Steel & Wire Co.	Steel Plant	Waukegan, Ill.
240	25	Tube cracked	Hammermill Paper Co.	Paper Mfg.	Erie, Pa.
241	25	Section of heating boiler cracked	F. A. Poth	Residence	Philadelphia, Pa.
242	26	Tube ruptured	Terminal Freezing & Heating Co.	Heating Refrig.	Baltimore, Md.
243	26	Blow-off pipe fails	Midland Chemical Co.	Chemical Works	Argo, Ill.
244	26	Boiler bulged and ruptured	Whitney Manufacturing Co.	Cotton Mfrs.	Whitney, S. C.
245	26	Boiler exploded	A. C. Shaner	Oil Drilling	Boliver, N. Y.
246	26	Boiler exploded		Oil Drilling	Ozark, Ark.
247	26	Five sections heating boiler cracked	Lauzon Furniture Co.	Furniture Mfg.	Gd Rapids, Mich.
248	27	Boiler exploded	B. F. Freymark	Machine Shop	St. Louis, Mo.
249	28	Boiler bulged and ruptured	L. T. Bornwasser Co.	Meat Packers	Louisville, Ky.
250	29	Tube ruptured	Alex Smith & Sons	Carpet Mfrs.	Yonkers, N. Y.
251	29	Boiler exploded		Flour Mill	Toledo, Ohio
252	30	Air tank exploded	Central Garage	Garage	Porterville, Cal.

MONTH OF MAY, 1920

No.	Day	Nature of Accident	Owner	Establishment	Location
253	1	Two sections heating boiler cracked	Alfred Brown	Apt. House	Worcester, Mass.
254	2	Section heating boiler cracked	Irvington Construction Co.	Theatre	New York, N. Y.
255	2	Flue of boiler fractured	Sister of Poor Hand Maids	Hospital	Mishawanka, Ind.
256	3	Crown sheet ruptured, 7 tubes pulled [loose	Southwestern Oil Corp.	Oil Refinery	Enid, Okla.
257	4	Blow-off pipe ruptured	Dallas Stm Laundry & Dye Wks.	Laun. & Dye Wks.	Dallas, Texas.
258	5	Two sections heating boiler cracked	St. Margaret's Memorial Hospital	Hospital	Pittsburgh, Pa.
259	6	Corrugated boiler furnace collapse	Western Brewery & Ice Co.	Ice & Cold Storage	Albuquerque, N. M.
260	6	Hot water boiler exploded	Ballard & Johnson	Restaurant	Peoria, Ill.
261	7	Accident to a boiler	McCombs Producing & Refn. Co.	Office Bldg.	E. St. Louis, Ill.
262	7	Section of heating boiler cracked	Atlantic Chain Corp.	Chain Mfgrs.	Long I. City, N. Y.
263	11	Bulged fire sheet	Original Min. Spring & Hotel Co.	Hotel	Oakwville, Ill.
264	11	Accident to a boiler	Swanson Packing Co.	Meat Packers	N. Sacra., Cal.
265	11	Pipe fitting burst	Lackawanna Steel Co.	Steel Plant	Lackawanna, N. Y.
266	12	Water heater exploded		Restaurant	Oakland, Cal.
267	12	Boiler exploded		Saw Mill	Clarksbg, W. Va.
268	12	Pipe fitting burst	King Paper Co.	Paper Mill	Kalamazoo, Mich.
269	14	Tube burst — town was without car service or light	Jackson St. Railway & Light Co.	Power Plant	Jackson, Tenn.
270		Tube ruptured	American Steel & Wire Co.	Steel Plant	Waukegan, Ill.
271	17	Accident to an economizer	Commonwealth Edison Co.	Power Station	Chicago, Ill.
272	17	Cracked header in water tube boiler	Barrett Mfg. Co.	Roofing Material	Elizabeth, N. J.

MONTH OF MAY, 1920 (Continued)

No.	Day	Nature of Accident	Killed	Injured	Concern	Business	Location
273	20	Furnace sheet of boiler bulged			Henneberger Ice Co.	Ice Plant	Mt. Carmel, N. Y.
274	20	Tube ruptured			Eastern Mfg. Co.	Paper Mill	So. Brewer, Me.
275	21	Hot water heater exploded			Emma Apartments	Apt. House	Peoria, Ill.
276	21	Tube failed and two headers cracked			Nekoosa-Edwards Paper Co.	Paper Mill	Pt. Edwards, Wis.
277	24	Section of heating boiler cracked			Meals Printing Co.	Printers	Gardner, Mass.
278	24	Tube ruptured		1	Washington Steel & Ordnance Co.	Steel Plant	Gresh'o M'r, D.C.
279	24	Fire sheet ruptured			Kingan & Co.	Meat Packers	Indianapolis, Ind.
280	25	Rendering tank exploded			Littlefield & Sons Co.	Rendering Works	Auburn, Me.
281	25	Tube burst			Standard Paper Co.	Paper Mill	Kalamazoo, Mich.
282	26	Accident to heating boiler		1	American Railway Express Co.	Express Co.	Worcester, Mass.
283	26	Boiler exploded; resulting fire did $75,000 damage		2	M. S. Richardson	Garage	Orchard, Col.
284	26	Boiler of locomotive exploded	2		Michigan Central R. R.	Railroad	Bay City, Mich.
285	26	Two tubes pulled out of drum			Grasselli Chemical Co.	Chemical Plant	Grasselli, Ind.
286	28	Boiler exploded		1	Jackson Blake	Saw Mill	Cameron, W. Va.
287	29	Tube burst			Standard Paper Co.	Paper Mill	Kalamazoo, Mich.
288	31	Tube burst			Alexander City Mills	Cotton Mills	Alexander City, Ala.

MONTH OF JUNE, 1920

No.	Day	Nature of Accident	Killed	Injured	Concern	Business	Location
289	1	Bagged firesheet			Marianna Ice & Cold Storage Co	Cold Storage	Marianna, Ark.
290	2	Boiler bagged and cracked			Lowman Brothers	Cotton Ginners	Staples, Texas.
291	4	Header in water tube boiler ruptured			Industrial Works	Construct'n Mach.	Bay City, Mich.
292	5	Boiler of locomotive exploded		1	Norfolk & Western Railroad	Railroad	Roanoke, Va.
293	5	Boiler exploded		1	John Glassick	Feed Mill	Cross Roads, Pa.
294	7	Hot water heating boiler cracked			Dwight State Company	Hotel	Springfield, Mass.
295	8	Cap on steam header blew off			St. Paul Sanitarium	Sanitarium	Dallas, Texas.
296	9	Tube ruptured			Nebraska Power Co.	Power Station	Omaha, Nebr.
297	10	Boiler exploded		1	Home Laundry Co.	Laundry	Devils Lake, N. D.
298	10	Water leg of boiler ruptured			Jacob Dold Packing Co.	Meat Packers	Wichita, Kans.
299	11	Header in water tube boiler cracked			Amer. Steel & Wire Co.	Steel Plant	Waukegan, Ill.
300	11	Tube ruptured			Hodenpyl-Hardy & Co.	Power Station	Flint, Mich.

No.	Day	Description	No.	Company	Type	Location
301	12	Boiler exploded		Gold Cheese Co.	Cheese Makers	Pound, Wis.
302	12	Four headers in one tube ruptured		H. P. Caiman & Sons	Canners	Bridgeville, Del.
303	13	Boiler of locomotive exploded		Baltimore & Ohio R. R.	Railroad	Washington, Pa.
304	13	Boiler of locomotive exploded	4	Chicago, Rock Island & Pacific	Railroad	Paxico, Kans.
305	18	Two tubes and three headers ruptured	2	Susquehanna Collieries Co.	Coal Mining	Shamokin, Pa.
306	19	Boiler stop valve cracked	1	Morris Ice Company	Ice Plant	Jackson, Miss.
307	19	Boiler exploded (See locomotive for October, 1920).				
308	20	Boiler exploded	1	Detroit Creamery Company	Creamery	Owasso, Mich.
309	21	Header in water tube boiler ruptured		Alabama Packing Company	Cold Storage	Acteco, Ala.
310	21	Two tubes pulled out	1	Monsanto Chemical Company	Chemical Plant	St. Louis, Mo.
311	23	Accident to steam valve		Kentucky & W. Va. Power Co.	Power Station	Lothair, Ky.
312	23	Boiler exploded	2	J. F. Scheberlie	Lumber Mill	Little Rock, Ark.
313	23	Boiler exploded		Bruce Hardwood Lumber Co.	Rock Drilling	Darlington, Wis.
314	23	Boiler on st'mship exploded. Vessel sank	1	Duquesne Steel Company	Steel Plant	Duquesne, Pa.
315	25	Boiler exploded	6	Steamer State of Washington	Steamship	Tongue Pt., Ore.
316	26	Tube ruptured		Steamtug Jerry	Steamtug	Staten Is., N. Y.
317	28	Boiler ruptured	2	H. C. Lytton	Store Building	Chicago, Ill.
318	29	Section of heating boiler cracked			Oil Drilling	Eldorado, Kan.
319	29	Section of heating boiler cracked	1	Fifty West Sixty-Seventh Street	Apt. House	New York, N. Y.
				W. A. Bradford	Office Bldg.	Quincy, Mass.

The Hartford Steam Boiler Inspection and Insurance Company

ABSTRACT OF STATEMENT, JANUARY 1, 1920.

Capital Stock, . . . $2,000,000.00.

ASSETS.

Cash in offices and banks	$390,221.07
Real Estate	90,000.00
Mortgage and collateral loans	1,426,250.00
Bonds and stocks	5,702,983.62
Premiums in course of collection	597,171.35
Interest accrued	107,590.44
Total assets	$8,314,216.48

LIABILITIES.

Reserve for unearned premiums		$3,715,903.48
Reserve for losses		175,539.16
Reserve for taxes and other contingencies		401,420.50
Capital stock	$2,000,000.00	
Surplus over all liabilities	2,021,353.34	
Surplus to Policy-holders		$4,021,353.34
Total liabilities		$8,314,216.48

CHARLES S. BLAKE, President.

FRANCIS B. ALLEN, Vice-President.   W. R. C. CORSON, Secretary.

L. F. MIDDLEBROOK, Assistant Secretary.

E. SIDNEY BERRY, Assistant Secretary.

S. F. JETER, Chief Engineer.

H. E. DART, Supt. Engineering Dept.

F. M. FITCH, Auditor.

J. J. GRAHAM, Supt. of Agencies.

BOARD OF DIRECTORS

ATWOOD COLLINS, President,
Security Trust Co., Hartford, Conn.

LUCIUS F. ROBINSON, Attorney,
Hartford, Conn.

JOHN O. ENDERS, President,
United States Bank, Hartford, Conn.

MORGAN B. BRAINARD,
Vice-Pres. and Treasurer, Ætna Life
Insurance Co., Hartford, Conn.

FRANCIS B. ALLEN, Vice-Pres., The
Hartford Steam Boiler Inspection and
Insurance Company.

CHARLES P. COOLEY,
Hartford, Conn.

FRANCIS T. MAXWELL, President,
The Hockanum Mills Company, Rockville, Conn.

HORACE B. CHENEY, Cheney Brothers
Silk Manufacturers, South Manchester,
Conn.

D. NEWTON BARNEY, Treasurer, The
Hartford Electric Light Co., Hartford,
Conn.

DR. GEORGE C. F. WILLIAMS, President and Treasurer, The Capewell
Horse Nail Co., Hartford, Conn.

JOSEPH R. ENSIGN, President, The
Ensign-Bickford Co., Simsbury, Conn.

EDWARD MILLIGAN, President,
The Phœnix Insurance Co., Hartford,
Conn.

EDWARD B. HATCH, President,
The Johns-Pratt Co., Hartford, Conn.

MORGAN G. BULKELEY, JR.,
Ass't Treas., Ætna Life Ins. Co.,
Hartford, Conn.

CHARLES S. BLAKE, President,
The Hartford Steam Boiler Inspection
and Insurance Co.

Incorporated 1866.

Charter Perpetual.

INSURES AGAINST LOSS FROM DAMAGE TO PROPERTY AND PERSONS, DUE TO BOILER OR FLYWHEEL EXPLOSIONS AND ENGINE BREAKAGE

Department.	Representatives.
ATLANTA, Ga.,	W. M. Francis, Manager.
1103-1106 Atlanta Trust Bldg.	C. R. Summers, Chief Inspector.
BALTIMORE, Md.,	Lawford & McKim, General Agents.
13-14-15 Abell Bldg.	James G. Reid, Chief Inspector.
BOSTON, Mass.,	Ward I. Cornell, Manager.
4 Liberty Sq., Cor. Water St.	Charles D. Noyes, Chief Inspector.
BRIDGEPORT, Ct.,	W. G. Lineburgh & Son, General Agents.
404-405 City Savings Bank Bldg.	E. Mason Parry, Chief Inspector.
CHICAGO, Ill.,	J. F. Criswell, Manager
209 West Jackson B'l'v'd	P. M. Murray, Ass't Manager.
	J. P. Morrison, Chief Inspector.
	J. T. Coleman, Ass't Chief Inspector
	C. W. Zimmer, Ass't Chief Inspector
CINCINNATI, Ohio,	W. E. Gleason, Manager.
First National Bank Bldg.	Walter Gerner, Chief Inspector.
CLEVELAND, Ohio,	H. A. Baumhart, Manager.
Leader Bldg.	L. T. Gregg, Chief Inspector.
DENVER, Colo.,	J. H. Chesnutt,
918-920 Gas & Electric Bldg.	Manager and Chief Inspector.
HARTFORD, Conn.,	F. H. Kenyon, General Agent.
56 Prospect St.	E. Mason Parry, Chief Inspector.
NEW ORLEANS, La.,	R. T. Burwell, Mgr. and Chief Inspector.
308 Canal Bank Bldg.	E. Unsworth, Ass't Chief Inspector.
NEW YORK, N. Y.	C. C. Gardiner, Manager.
100 William St.	Joseph H. McNeill, Chief Inspector.
	A. E. Bonnett, Ass't Chief Inspector.
PHILADELPHIA, Pa.,	A. S. Wickham, Manager.
142 South Fourth St.	Wm. J. Farran, Consulting Engineer.
	S. B. Adams, Chief Inspector.
PITTSBURGH, Pa.,	Geo. S. Reynolds, Manager.
1807-8-9-10 Arrott Bldg.	J. A. Snyder, Chief Inspector.
PORTLAND, Ore.,	McCargar, Bates & Lively,
306 Yeon Bldg.	General Agents.
	C. B. Paddock, Chief Inspector.
SAN FRANCISCO, Cal.,	H. R. Mann & Co., General Agents.
339-341 Sansome St.	J. B. Warner, Chief Inspector.
ST. LOUIS, Mo.,	C. D. Ashroft, Manager.
319 North Fourth St.	Eugene Webb, Chief Inspector.
TORONTO, Canada,	H. N. Roberts, President Boiler Inspection
Continental Life Bldg.	and Insurance Company of Canada.

AUTOMATICS

In "Power"

By Rufus T. Strohm.

There are scads of strange devices in our power plants today,
Made to put an end to trouble and to keep the jinx away,
While a lot of them are fashioned with the pleasing end in view
Of avoiding heavy labor that would fall to me and you;
But although they're almost human in the way they do their share,
They require close inspection and they need continued care;
So some accidents will happen and some trials will remain,
For **we** haven't found the fellow with **an automatic brain.**

You can use **a** warning whistle **of** the kind that starts to blow
When the water in the boiler gets too high or falls too low,
But your hair will turn to silver while you wait to hear it wail,
If the water-column piping isn't free from sludge and scale;
And unless the water tender is alive and on the jump,
It will never halt disaster though it blare like Gabriel's trump;
For **a** boiler can't forevermore withstand **an overstrain,**
And we haven't found **the** fellow with **an automatic brain.**

You may furnish every engine with a form of safety stop,
With the object of protecting all the workers in the shop,
But you've got to clean and oil it, or its joints will rust and grip,
And you'll find it out of order when you're praying it will trip.
So, you see the risks you're running and the chances that you take
If the chief and his assistants aren't keen and wide awake;
Troubles thrive where men are careless, foolish, ignorant or vain,
Since we haven't found the fellow with an automatic brain.

Thus with stokers, lubricators, non-return valves and the like —
They are mighty clever helpers, but they're sure to go on strike
If, when you have made them part of your equipment, you proceed
To forget you ever bought them and ignore the care they need.
You will find that they'll protect you, and your drudgery will shrink,
But they're only brass and iron, and they weren't made to think;
So you needn't look for absolute security to reign
Till **the race** brings forth the fellow with an automatic **brain.**

The following stanza is added with apologies to the Author.

But when the hand upon the pressure gauge swings 'round against the stop
And the safety valve relied upon gets stuck and will not pop.
If the safety stop won't function and the engine goes awry
So that pieces of the flywheel upward sail unto the sky,
Or if the boiler gets to foaming and the engine gets a shot
Of some water in the cylinder and things go all to pot,
Then's the time a Hartford Policy will save you from the pains
Of a bank-roll **cut to** pieces by non-automatic brains.

The Locomotive

DEVOTED TO POWER PLANT PROTECTION
PUBLISHED QUARTERLY

VOL. XXXIII.　　HARTFORD, CONN., APRIL, 1921.　　No. 6.

BOILER EXPLOSION AT SAVANNAH, GEORGIA.

Boiler Explosion at Savannah, Georgia.

THE year 1920 closed in an unfortunate way for the firm of G. H. Tilton and Sons of Savannah, Georgia. About the middle of the afternoon of December 29th, while everything was running smoothly and the year's work was nearing its close, one of the boilers in the mill of this concern exploded with terrific violence and serious damage was done to the plant. A general view of the ruins appears on the front cover of this issue of The Locomotive and, as will be seen by reference to that picture, the wreckage was

spread over a large area. The boiler room was completely demolished as was also a brick warehouse nearby. The boiler, in exploding, ripped off one course or sheet which landed some distance away from the mill. The rest of the boiler was hurled through the picker room, adjoining the boiler room, and was brought to rest by the wall of the main mill building but only after it had torn a fair sized hole in that wall. In the picture above, the boiler is shown where it landed after its flight. A piece of pipe three inches in diameter and about eighteen feet long was thrown through the air and came down through the roof of a dwelling fifty yards away.

The property loss amounted to about $40,000. In addition to the property damage, however, and for which money cannot adequately pay, was the injury to the two firemen at the plant. One of these men was so seriously injured that he was not expected to live.

The cause of the explosion is not definitely known though it would appear, from the evidence we have at hand, to have been another case of a lap seam crack, reference to which was made in the last issue of The Locomotive.

Flywheel Accident at The Kentucky Utilities Company.

BUSINESS which required light or power was recently brought to a standstill in Richmond, Kentucky, when a flywheel on an engine at the plant of the Kentucky Utilities Company exploded and so damaged the building and equipment that the plant was forced to shut down. The illustration herewith gives a view of the wrecked engine and some of the damage that was done to the building.

FLYWHEEL EXPLOSION.
KENTUCKY UTILITIES CO.

Aside from the economic loss produced by the interference with business, the accident carries with it the sad story of the loss of a life. Driven by its tremendous momentum, part of the wheel ripped out through the roof of the plant and buried itself in the ground almost half a mile away. In its flight it crashed through a house.

which it entered through the roof, passing through the upper floor and coming out through the side wall. As it did so, it struck and instantly killed a child and injured a young woman who lay asleep directly in the path of the flying missle.

The accident came without warning. The engineer, but a few moments before, had been oiling the engine and had just stepped out to the boiler room when the crash came. So far as we have been able to learn, no explanation has been obtained as to the cause of the explosion.

Engine Foundations.

FOUNDATIONS for engines are necessary for several reasons. The engine must, first of all, be held in alignment — alignment of its own parts and with the external machinery that it drives. The foundation must distribute the weight of the engine and also its own weight to a ground area sufficiently great to avoid exceeding the bearing power of the soil. A third function that the foundation must perform is to absorb or dampen the vibrations of the machinery resting upon it so that these vibrations will not be exaggerated in the machine itself nor transmitted to the surrounding machinery and buildings.

Fifteen or twenty years ago it was common practice to build engine foundations of brick or stone. It is modern practice, however, to use concrete as this material is not only more convenient but also is better adapted to the requirements of a foundation. Practically all concrete foundations of ordinary size may be made monolithic — that is to say, in one solid block without joints. Even with foundations of great size the joints will be few and with proper precautions the bond at these joints may be made practically as strong as any other part of the structure.

It is somewhat difficult to state a rule, that will cover all cases, as to the proper size of a foundation. For the weight to be from four to five times that of the engine is in accordance with average practice. The factor of primary importance, however, is the ability to remain fixed in position. Settling or sinking of the foundation cannot be absolutely avoided but it can be reduced to a negligible amount and in most cases to an amount so small that it cannot be measured by any ordinary means. Whatever settling does occur should be uniform. To provide adequate bearing for the foundation, the structure should

be proportioned in accordance with the bearing power of the soil as
given in Table I. Knowing the weight of the engine and the approximate weight of the foundation, it is easy to calculate the area that
the foundation should **cover**. In this calculation the weight of con-
crete should be taken as 150 lbs. per cubic foot.

TABLE 1.[1]

BEARING POWER OF SOILS AND ROCK.

Ledge rock .... 36 tons per sq. foot
Hard pan .... 8 " " " "
Gravel .... 5 " " " "
Clean sand .... 4 " " " "
Dry clay .... 3 " " " "
Wet clay .... 2 " " " "
Loam .... 1 ton " " "

The **foundation** must, of course, be deep enough to be entirely
free from disturbance by frost. If bed rock can be reached at not
too great a **depth**, it would be well to excavate to that depth. A
bearing on **bed rock** is not always desirable, however, as will be
shown later. In some cases the nature of the soil may be such that
the area necessary
to support the
weight will not be
great enough to
bring the bearing
pressure on the
soil within the
safe limit without
excavating to an
extreme depth. In
such cases it will
be best to exca-
vate to a moder-
ate depth and
then to construct
a mat, or sub-
foundation, as
shown in Fig. 1.

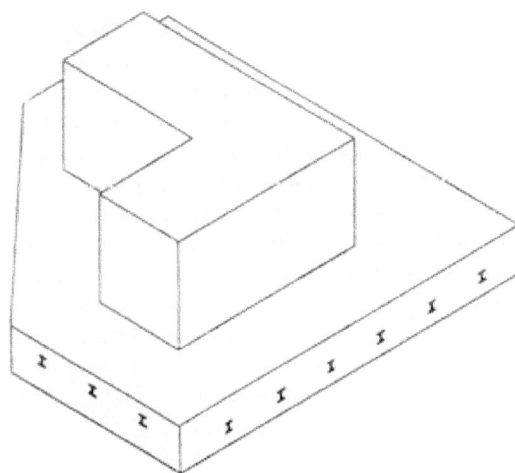

FIG. 1

[1] Taylor & Thompson "Concrete, Plain and Reinforced."

which will distribute the load over a greater area. In any ordinary case it would probably not be advisable to go deeper than twelve feet to reach a bearing soil, for if an adequate bearing stratum is not reached within this distance it would be better to construct a mat or to sink piling and then place the foundation upon a large mat or cap built upon this piling. This mat should be well reinforced with bar iron. The regular reinforcing steel is, of course, quite acceptable for this purpose although a few lengths of railroad iron laid at right angles to each other is an excellent reinforcement in a construction such as this. The rails should be cleaned of oil and grease before using.

There have been installations in which the foundations of the walls of the buildings have been placed on bed rock and vibration has developed in the building when the machinery also had its foundations resting on this rock. In such cases, when the rock lies so near the surface that it is difficult to avoid placing the machinery foundations upon it, it is well to blast an opening in the rock so that a sand cushion may be placed between the engine foundation and the rock as illustrated in Fig. 2. A box or retaining wall of concrete as shown in Fig. 3, may also be used to accomplish this purpose. The necessary spread or area of foundation is, of course, based on the allowable bearing pressure on the rock.

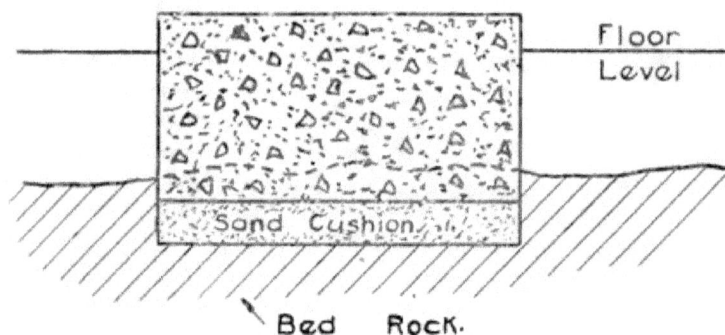

FIG. 2.

When the soil will permit it, the excavation itself may provide the form for pouring the concrete. If the ground is not firm enough for this, wooden forms must be built. Battered or sloping slides may be used when it is desired that the foundation have additional spread or bearing area. It is more convenient, however, and it is just as satisfactory to use vertical sides and secure additional bearing surface

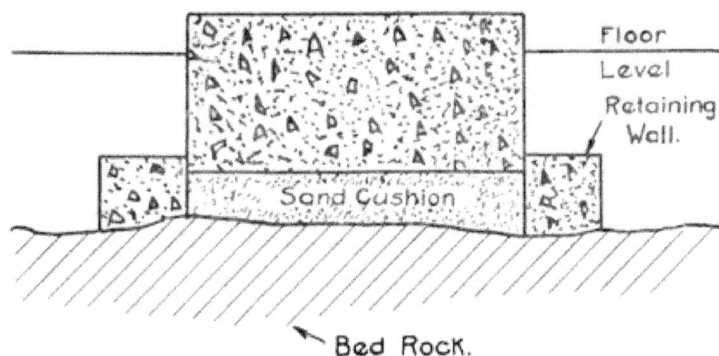

Fig. 3.

by the use of a mat or sub-foundation. The plan of the foundation follows, roughly, the outline of the base of the engine. When there is an outboard bearing the design should be somewhat as shown in Fig. 4 in which the foundation or pier under the outboard bearing is tied to the rest of the structure by a slab of concrete. With such a construction some thought should be given to the weight imposed upon the foundation by the outboard-bearing pedestal and the weight of the parts that such bearing carries. Even though the total area of the foundation may be suffi-ciently great to support the total weight without exceeding the safe bearing pressure, it is quite pos-sible that the load may become highly concentrated at certain points such as that under the out-board bearing. In such a case it would be well to spread the base of the foundation under the ped-estal by sloping the walls or by a

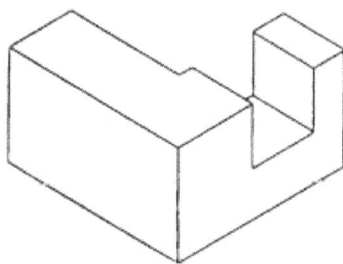

Fig. 4.

sub-foundation to bring the pressure within the allowable limits.

In constructing the form for the foundation it is well to use green or unseasoned lumber because such lumber does not have as great a tendency to warp out of shape in the presence of water as it would if it were seasoned. Spruce is an excellent wood to use and fir is also very satisfactory. The lumber for foundations below ground may be one-inch boards. It should be borne in mind that the water in concrete is necessary for the proper formation of the cement crystals, and that the forms should be as water tight as it is practicable to make them so that the water will be retained within the concrete until it hardens. For this reason the edges of the boards of the forms should be joined. Even better than the joining by square edges is the construction in which the edges are beveled as indicated in Fig. 5. When the wood becomes wet and swells, these edges are forced together into a water-tight joint. This construction is favored, by many, above that in which tongued and grooved boards are used.

The battens or studs, to which the boards of the form are fastened, should never be less than 2" x 4" nor spaced more than two feet apart. The battens should be braced in position so that the form will hold its shape and position, and the braces should be securely fixed at their outer ends. When it is impossible to provide a sufficient number of braces outside of the form, bolts or wires may be run through the foundation itself to tie the opposite sides of the form together.

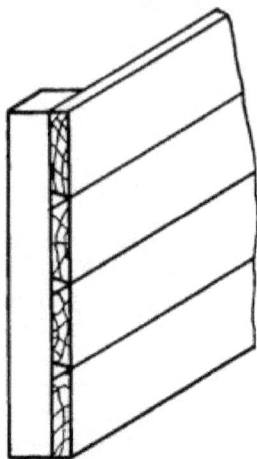

FIG. 5.

When foundations are entirely below ground there is no necessity for a smooth finish nor does it matter if the forms bulge slightly so that the sides of the foundation do not form a plane surface. With foundations that are visible, however, attention must be given to these details. In such cases care must be taken to see that the surfaces of the boards in contact with the concrete are planed and smooth unless, of course, a cement plaster finish is to be given later. It is well also to use 1½" or 2" planking and to pay some extra attention to the bracing so that true surfaces will be obtained. The obtaining of a good finish will be discussed a little later.

When the sides of the foundation form have been constructed they may be placed in position and fastened together as shown in Fig. 6. The sides should be made square with each other and then braced in that position with boards laid diagonally across and fastened to the top of the form. Lines should be run across the top of the form to indicate the center lines of the shaft and the cylinder, and the form should then be placed accurately in position by reference to these lines and securely held in this position. The engine, when it is set on the foundation, is not to rest upon the concrete itself but will have, between it and the concrete, a bed of cement grout (equal parts of sand and cement, about one-half inch thick. This method of setting in grout insures the accurate placing of the engine and at the same time provides a full and even bearing for the engine bed-plate. When placing the foundation forms, therefore, it is well to have the top edge of the sides straight and level and located one-half inch below the level that the bed plate of the engine is to assume. The form is then filled with the concrete to the point of overflowing.

The next step is to support the holding - down or foundation bolts in the proper position in the foundation opening. To do this a template, such as that shown in Fig. 7, on page 170, is necessary. The center lines of the engine shaft and of the cylinder are carefully laid out at right angles to each other on the template and

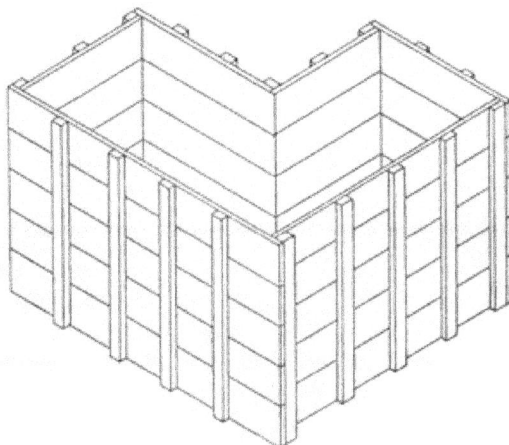

Fig. 6.

the bolt holes are then located from these center lines. The location of the bolt holes in the bed plate of the engine is secured either by direct measurement or from the makers. The length of the foundation bolts should be, if possible, thirty or more times their diameter. That is to say, a one-inch bolt should be thirty inches or more in length. This is the length of the bolt that is within the foundation and to it should be added the necessary length to extend above the foundation to pass through the grouting, the bed plate of the machine, and the nut and washer above. There is considerable advantage in having the bolts extend an inch or two above the level theoretically required, for inaccuracies of setting or measurement may cause this extra length to be of considerable value. If, after the engine is finally set, the bolts extend beyond the nuts, the protruding ends may readily be removed with a hack-saw.

The holding-down bolts should not be rigidly fastened in the foundation because, for one reason, the bolt holes in the bed-plate

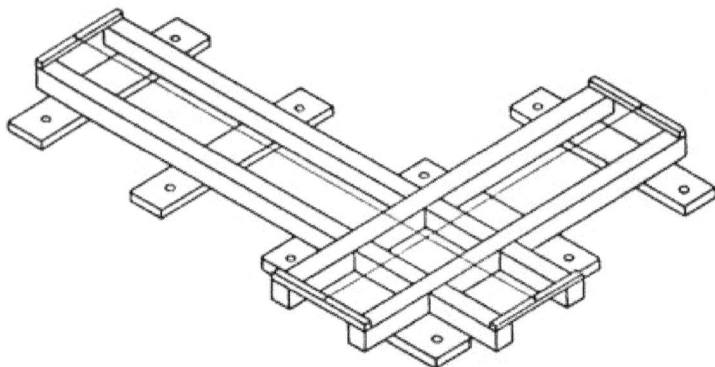

Fig. 7.

casting may not agree in position with the design. Then too, there is always the possibility of error in the layout of the template. The engine, after being placed on the foundation and located by fixed foundation bolts, may not line up with the machinery it is to drive, even though the foundation and template were carefully located. It is best, therefore, to use some method of holding the bolts in the foundation that will provide a fair degree of freedom so that they can be entered in the bed-plate holes and the engine shifted slightly to bring it into alignment. This can be accomplished by the method of Fig. 8 which shows a piece of pipe, whose internal diameter is from one to two inches greater than that of the bolt, suspended by the bolt from the template. This piece of pipe should be centered with the pipe at the top and bottom by means of centering-disks as shown. These disks may be made of wood or metal. It is very convenient to have the lower centering-disk cast upon the upper face of the foundation washer. Wooden wedges or even paper or waste may be used at the top of the pipe where the centering of the bolt in the pipe can readily be observed. Wooden boxes may be used, of course, instead of the pipe but when used they should be made wet before the concrete is poured. This will prevent the wood absorbing moisture from the concrete — which would prevent the formation of a hard concrete — and it is of value also in that the boxes will later dry out, shrink, and thus be easy to remove from the foundation.

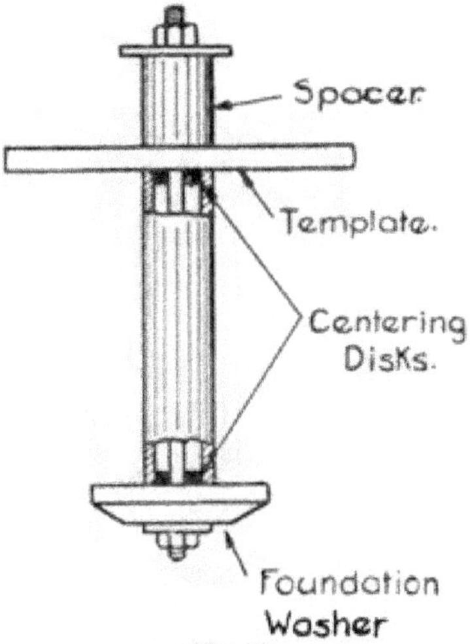

Fig. 8.

The washers for use on the lower end of the holding-down bolts may be made of cast iron and should be of good size. Proportions for these washers are given in Table II. The nut on the lower end of the bolt will be held from turning if the concrete, when poured, is puddled or rodded around the lower end of the bolt although it is advisable to have the bolt extend through the lower nut for about twelve inches and to have this extension bent out of line so as to act as a lever. Ridges may also be cast on the foundation washer to prevent turning of the nut.

In many cases, especially those in which the foundation bolts extend for some distance above the top of the foundation, it will be convenient to have the bolts so installed that they may be removed.

Table II.

Proportions of Cast Iron Foundation-Washers.

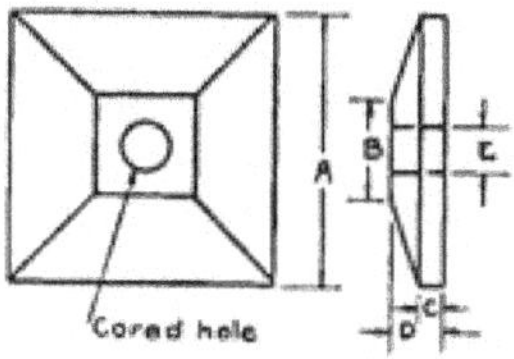

Dia of Bolt	A	B	C	D	E
1"	6½"	2½"	⅝"	1¼"	1¼"
1½"	9½"	3½"	11/16"	2⅞"	1¾"
2"	13"	4½"	1¼"	2½"	2¼"
2½"	16"	5"	2⅝"	3¼"	2¾"

and for this purpose the foundation should be constructed with
pockets to hold the lower washer and nut. The details of such a
construction are shown in Fig. 9. When this plan is followed, the
bolt holes may best be formed by boxes which are held in place by
cross braces or stays at the top and by the form for the pockets at
the bottom. This construction for the
bolt holes is of great advantage when
placing heavy machinery. There are no
bolts to interfere with the moving of
the machine and it can easily be rolled
into position and the bolts dropped in
later on. When the bolts are not removable a great amount of time is often
lost by the necessity of elevating the
machine above them and, furthermore,
damage to them may very readily occur.

The best mixture for making concrete for foundations is what is known
as the 1 :2 :4 mixture or, in other words,
one part cement, two parts sand, and four
parts crushed stone (all parts by volume. This mixture forms a dense, hard
concrete. When a foundation is very

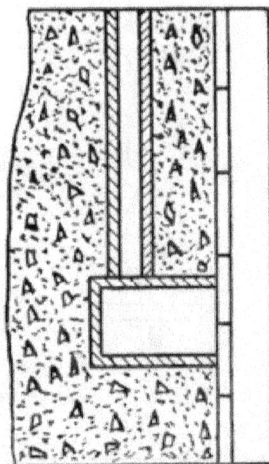

Fig. 9.

large the lower part of it may be built of a 1 :2½ :5 or 1 :3 :6 mixture
with a top of the 1 :2 :4 mixture about two feet thick.

The cement is the binder for the whole structure and care should
be used to see that a good grade of Portland cement is obtained. The
sand should be clean and sharp and it is better for it to be composed
of a mixture of large and small grains than for it to be all of uniform size. The stone may be used as it comes from the crusher after
the dust and screenings have been removed. The stone should all be
of such a size, however, that it will readily pass through a 2½" ring.

A power driven concrete mixer is, of course, the most convenient
means of mixing and in large foundations is practically an absolute
necessity as all concrete foundations should be carried to completion
without interruption when the pouring has started. The power mixer
has also the advantage of giving a thoroughly mixed concrete. There
is no serious objection, however, to the hand method of mixing if
the pouring of the concrete can be completed within a reasonable time.

When hand mixing is employed, a mixing platform about fifteen
feet square should be built. One inch boards supported on joists

which are spaced three feet apart will, in most cases, be satisfactory. A 2" x 3" strip running around the edge of the platform is useful in preventing the waste of materials. For the purpose of measuring the materials, a bottomless box with a capacity of four cubic feet will be of advantage. It will be found convenient to have this box made approximately to the dimensions of 18" wide x 32" long x 12" deep.

In making up a batch of concrete the measuring box is placed on the mixing board, filled to half its depth with sand and then removed. One standard size bag (approximately 1 cubic foot) of cement is added to the sand and the two are thoroughly mixed together while dry. The mixture is then leveled off, the measuring box placed on it and a full measure, or four cubic feet, of the broken stone added. The three ingredients are then thoroughly mixed while dry, by turning with shovels and by raking. When thoroughly mixed, the water is added gradually while the mixture is again turned and re-turned and raked and re-raked. Sufficient water is added to give a consistency of thick cream. Of course, double or triple the above quantity may be mixed at once if it is found convenient to do so.

When the concrete is well mixed, it may be placed in the form and evenly distributed. The surface should be kept approximately level at all times as the work progresses, and the concrete should be tamped and worked with rods or bars into all corners of the form. Care should be used about the foundation bolt casings so that they will not be thrown out of a vertical position. It is well, as an added precaution, to hold the lower end of these casings in position by tie wires running from casing to casing and from casing to foundation form.

At least one week — a longer time is advisable — should elapse between the pouring of the concrete and the placing of the engine upon it. During this time it is well to keep the concrete wet to assist the setting because cement develops its strength by the formation of crystals and water is essential for this action. Once the crystallization is started it should continue to completion, for if interrupted a weak concrete will result. If for any reason the pouring of the concrete is interrupted before completion of the work, the surface of the foundation should be kept wet by placing water-soaked bags upon it. On the other hand it should not be mixed too wet, for excessive water in the mixture will wash the sand-and-cement mortar away from the stones and weakness will result. It might be mentioned here that if the pouring is interrupted it may be continued later upon the old surface if that surface is not too smooth and has been kept

wet and free from dust. Otherwise the old surface should be
" picked " until rough and then thoroughly washed, and this procedure
should also be followed just before the engine is to be placed on the
foundation.

The engine, after being placed in an approximately correct position on the foundation, must be levelled and aligned with the machinery it is to drive. For the purpose of leveling, some iron wedges
with a rather small taper must be provided. These are laid on the
foundation so that when the engine is placed the wedges will be distributed around the edge of the plate. Enough wedges should be
provided to enable them to be spaced not more than six feet apart
and, in any case, a wedge should be placed close to each foundation bolt. The machine is to be made level with reference to the
shaft and to the cylinder and cross head slide. This is easily determined by placing a machinist's level on a machined surface of the
parts mentioned. The crosshead slide is convenient for the one direction. For leveling in the other direction the level must be carefully
placed on the top of the shaft at a place where the shaft is perfectly cylindrical and does not taper. By driving on the wedges with
a heavy hammer or sledge the engine may be brought to a perfectly
level position in both directions. At the same time that the machine
is made level it must, of course, also be brought into alignment with
the line shaft or other machinery that it is to drive. When the engine has been made level and in alignment, the nuts on the foundation bolts should be tightened and each wedge should be struck a
light blow to set it up firmly without disturbing the setting.

When the setting of the engine has been accomplished, it should
be resting with a space of about one-half inch between the bottom of its
bed-plate and the top of the foundation. A dam is then built about
the base of the engine with timbers, about 4 x 4" in size, laid two
inches or more away from the sides of the base. Grout, composed of
equal parts of Portland cement and clean sand mixed with water to
about the consistency of cream, is poured within this dam so as to
flow under the engine. Before pouring, however, the opening in the
foundation around the bolts should be plugged at the top with waste
or paper, so that the grout cannot run into and fill these holes. The
mixture, when poured, should be stirred and worked in and under
the engine base to remove all air bubbles and to insure complete contact of the grout with the foundation and engine bed-plate. Enough
should be poured to fill the dam to the point of overflowing as this
will insure its reaching all parts that are to be imbedded. After the

grout has set a few hours the dam may be removed and the surface smoothed with a trowel. Do not, however, remove the wedges until about a week's time has elapsed. They should always be removed and the bolts made tight before the engine is put into service but care should be used in doing this so that the setting of the engine is not disturbed.

If a smooth finish is desired the foundation may be plastered with a mortar composed of equal parts of sand and cement although it is sometimes difficult to secure a good bond between the concrete and the cement plaster. A better plan is to use a little extra care in the construction of the forms so that the surfaces of the boards are clean, smooth, and free from knots. Then by plastering the inside of the form a few inches in height at a time and just before a batch of concrete is poured, the plaster, which should be about $\frac{1}{4}''$ thick, and the concrete will be bonded while still wet. As a final finish and to prevent any oil permeating the structure and thereby weakening it, a coat of paint may be given the foundation when it has become thoroughly dry.

The Petroleum Situation.

THE Drake Well, which "struck" oil on Watson's Flats near Titusville, Penn., on August 28, 1858, marks what may be called the beginning of our petroleum oil industry. "Rock oil" had been produced prior to this date by digging shallow wells wherein the oil floated to the surface of the water and was then dipped up with blankets. Coal-oil, distilled from bituminous coal, was also being produced and, while not in extensive use, had awakened the people to the advantages of the oil lamp over the old-fashioned candle. Kerosene, obtained by distillation from crude petroleum, found therefore, a ready market. In fact, in the early days the demand for kerosene was the spur to the development of the oil fields. Benzine and gasoline were at that time considered as waste and were often physically destroyed. Because of its dilution by the lighter products of distillation, kerosene used in oil lamps was often the cause of serious accidents. Nowadays we complain of too much kerosene in our gasoline.

It has been said that the units of mechanical power per capita are a measure of our progress in civilization. Certainly there is no comparison between the power produced today and that used before the Civil War. In the olden days we could successfully lubricate our machinery with animal and vegetable oils and what little mineral oils were available. Today we are practically totally dependent upon petroleum for our supply of lubricating oils. In those earlier days crude oil and fuel oil were little used in industrial furnaces either for the production of steam or for the numerous purposes to which they are now applied such as heat treating furnaces, retorts, and kilns. Approximately one-half of the petroleum recovered from the earth is now being consumed in the direct production of heat.

Though we little realize it, we are in almost daily contact with the products of petroleum. Dyes for our clothes, flavoring extracts for our foods, chemicals and drugs in formidable array, all are obtained from this mineral resource. It has been estimated that there are eight million automobiles in use in the United States and that the figure may soon reach twelve million. The motor truck industry is growing by leaps and bounds, the farm tractor is fast becoming indispensable to agriculture, and the airplane, developed to a remarkable degree during the World War, is also finding its place of service. As rapidly as the crude oil is removed from the ground, new uses are found for its products.

Some idea of the enormous size of the petroleum industry may be gained from the statement that during the year 1920 we removed from the oil fields of this country 531,186,000 barrels of oil. To visualize it, think of the picture that has been drawn that this is equal to the flow of Niagara for over three hours. The question should naturally arise in the mind of anyone who gives thought to the future — how long will our petroleum resources continue to supply us at this rate?

The United States Geological Survey has made careful and reliable estimates of the amount of oil that has been removed from and that still remains in the ground. Since the drilling of the Drake Well in 1858 over five billion barrels of oil have been consumed and we have remaining in the ground and recoverable by present methods of production approximately seven billion barrels more. At the present rate of consumption seven billions reserve supply will last us about seventeen and one-half years. *At our present rate of consumption* — and every year we are finding new uses for the products of petroleum. Of course the inexorable law of supply and demand will step in before the expiration of that period so that actually our use of petroleum must gradually taper off.

It is estimated that by present methods we are recovering less than one-quarter and in many cases as little as one-tenth of the oil in the strata pierced by the wells. By improved methods there is promise of a greatly increased flow so that our total recoverable oil may be doubled, leaving us nineteen billion barrels available. Using our present yearly rate of consumption, we would reach the end in about fifty years. But even that is within the lifetime of the next generation.

Looking abroad, estimates indicate that the world's supply of petroleum amounts to about forty-three billion barrels. Of this, the United States has about one-sixth (based on the seven billion known to be recoverable). The world in general is increasing its consumption of oil just as is this country, so we cannot look to foreign fields for a fully adequate supply. It might be stated, in this connection, that in 1920 we imported 106,175,000 barrels of oil which is more than double the imports of 1919 and almost five times greater than the imports in 1913. It is true that one enormous resource, the oil bearing shales of Utah, Colorado and Wyoming, is yet to be developed and may solve many of our difficulties but " counting your chickens before they are hatched " is a poor policy and, furthermore, it must be realized that the labor question involved in the development of these oil shales is a tremendous one.

Dr. George Otis Smith, Director of the U. S. Geological Survey, presents the problem of the future supply of energy in a most interesting way. In comparing the energy resource of water power, coal, and oil he says:

" If we take fifty million horsepower as an average figure for the potential water power of the United States, without storage, we find that, if fully developed and if used at the average load factor of today, our rivers and streams would just about meet the country's present needs and would supply that amount of power for all time : moreover, with storage and an improved load factor they could provide a considerably increased output of energy to meet the growing demand.

" If, however, we should put the whole burden on our coal mines, not using even the water power now used, we would find that by adopting the best steam practice of today the present power requirements of this country could be met with coal for 57,000 years, although we know that long before the end of that period the greater depth of the coal mines and their increased distance from market would alone create power demands for mining and transportation, that would considerably cut down the amount of power available for other uses.

" We measure the petroleum wealth of the United States by billions of barrels — about five billions already produced in the last sixty years, and about seven billions left for the future. Again adopting the best steam practice of today in public-utility stations of Texas and California — a little less than thirty-two barrels to the horsepower-year — and trying to carry the whole power load of the country with oil alone, we find that the oil reserves of the United States, although measured by billions of barrels, would last only nine years and three months. Without allowing for the fact that steam raising for power is only one of the many uses of coal, these two figures — 57,000 years and 9¼ years — are sufficiently impressive to make us fairly receptive to the general truth of Mr. Eckel's statement in his recent book " Coal, Iron and War!" " ' We have just as much real chance of replacing coal by oil as we have of finding enough gold to use it in place of steel.' " These are only comparative estimates of the energy supplies which may be tapped for the use of our own citizens."

Petroleum, in the civilized world of today, is a commodity of far reaching importance. We find in isolated districts, where electricity is not available, that kerosene is the only adequate source of light. Our dependence on the automobile can no longer be looked upon as an extravagance but as a daily necessity. Lubricating oil must be had for the wheels of industry to turn and no adequate source other than petroleum is in sight. In many sections industries wholly dependent upon oil for fuel have been developed. Truly the petroleum situation is a serious one and demands our attention.

The enormous increase in demand for oil as fuel has arisen because of the ease of handling the oil and the difficulty, at times, of securing coal. Crude oil, with its content of benzine, gasolene, kerosene and the lighter lubricating oils, is hardly thought of for use in steam raising. The portion of the petroleum oil that is left after the lighter distillates have been removed, and which is commonly known as fuel oil, may be considered, in a broad way, as a solution of asphalts and waxes in lubricating oils. It would appear therefore that this fuel oil, like crude oil, may serve a higher purpose than that of steam raising.

Does it not seem that, since coal is what might be called a single purpose commodity in that it serves almost solely for the production of heat, and since petroleum is a multiple purpose commodity capable of filling wants which coal cannot fill, we therefore should look to the diverting of oil to its special uses? Oil may be abundant now but what of future generations? It is of course quite true that new oil fields may be opened up and also that methods of recovery far superior to any now in use may be developed. Some day, however, the supply must be exhausted. Should we hasten that time to the detriment to posterity?

Engine Wrecked by Excessive Water.

By Inspector R. Downie.

I N the last issue of The Locomotive mention was made, under the heading of " Excessive Water as the Cause of Engine Breakage," of the possibility of an engine being wrecked by water drawn into the cylinder from the exhaust pipe. An accident of this nature occurred recently at Utica, New York, and serious damage was done to the engine.

According to the engineer in charge, the engine was started up one Monday morning and, soon after it reached full speed, was wrecked. Several of the cylinder head bolts broke and allowed the cylinder head joint to leak enough to release some of the pressure although the cylinder head itself was not blown off. But before the pressure had been thus relieved, four of the six main-bearing-cap bolts were broken and the other two were elongated. The connecting-rod was bent at both ends although not so badly damaged but that it was possible to straighten and use it again. The crank-shaft, which was six inches in diameter, and the piston-rod were so badly bent that it was necessary to replace them with new parts.

We have been advised that the six-inch exhaust line on this engine was provided with but a ½" drain pipe. When this small drain pipe became stopped up, as it very likely did, considerable condensation accumulated in the exhaust pipe between the time of shutting down on Saturday and of starting on Monday morning. As soon as the engine reached normal speed with no load the governor cut down the steam admission to the cylinder to a very small amount, the cold cylinder walls readily produced a vacuum in the engine, and the water was thereby drawn into the cylinder and caused the damage. It hardly seems probable that the water reached the engine from the boiler as the water was well below the normal level at the time of the accident. To guard against a repetition of the accident, however, the exhaust line has since been provided with a larger drain pipe.

Engines may very often be subjected to any one of a number of conditions that will cause a sudden breakdown. Protection against the financial loss that is so brought about may, however, be secured by having the engine covered by a Hartford Engine Breakage Policy.

The Locomotive

Devoted to Power Plant Protection

Published Quarterly

Wm. D. Halsey, Editor.

HARTFORD, APRIL, 1921.

Single copies can be obtained free by calling at any of the company's agencies.
Subscription price 50 cents per year when mailed from this office.
Recent bound volumes one dollar each. Earlier ones two dollars.
Reprinting matter from this paper is permitted if credited to
The Locomotive of the Hartford Steam Boiler I. & I. Co.

TO be given a " square deal," machinery of every description should rest upon a firm foundation. In no case does this apply with greater force than in that of the steam engine. Although the inertia of the reciprocating parts is well balanced in a properly designed engine, yet there must inevitably be a varying load imposed upon the area on which the engine rests. A varying load is often more harmful than a steady pressure and inadequate support under such conditions must result in misalignment of the parts and, sooner or later, undue wear will develop even if actual breakage does not occur. This is particularly true in an engine covering a large area with the loads concentrated upon a relatively small number of bearing points.

We would therefore call particular attention to the article entitled, "Engine Foundations" which appears in this issue of The Locomotive. Not all the points involved in the design and construction of a foundation have been covered but we believe that the information given, together with some thought to local conditions, will be adequate for any ordinary installation. In any case, play safe. The added years of life that are given an engine by providing a well designed and well constructed foundation are well worth the extra cost.

OBITUARY.

Mr. Edward B. Hatch, a director of The Hartford Steam Boiler Inspection and Insurance Company, died on February 18th at his home in Hartford after a brief illness. This wholly unexpected termination of an exceptionally active life came as a great shock to a wide circle of friends and associates not only in the community in which he lived but also throughout the country generally. Mr. Hatch was well known throughout the business world and was highly regarded for the qualities which had made him successful in the industries under his care and also because he brought to every business relationship the qualities which distinguish a gentleman and the straightforward, honest dealing of a stern Christian character. After graduating from Trinity College, Mr. Hatch entered the employ of the Johns-Pratt Company of Hartford, which was then a young enterprise in a novel and untried line of business. He entered into his work there with the energy and devotion which characterized his business connections all his life and early was given the responsible direction of its affairs, to later become its President, a position which he held until his death. The Johns-Pratt Company is to-day one of Hartford's largest and most prosperous industries. Its present success stands as a monument to the ability and energy of Mr. Hatch's life.

Mr. Hatch was actively connected as officer or director with many other industrial and financial corporations both in Hartford and in other sections of the country. In his own city he took great interest in civic affairs. For several terms he served as a member of the Board of Water Commissioners of Hartford at a time when important enlargement of the water supply presented many serious problems for consideration, and it is recognized that their successful solution was in large part due to the constructive study and judgment which he gave in that service.

Mr. Hatch was elected a director of The Hartford Steam Boiler Inspection and Insurance Company in 1915 and served continuously from that date. He brought to the deliberations of the Board the viewpoint of a man of affairs and broad experience in the business world. As a manufacturer he knew the needs and requirements of that large class of business men whom our Company especially serves, and he was in position to advise how that class could best be served by us. Our Company has indeed sustained a great loss in his death. The Board of Directors of this Company, in a desire

to express their sorrow and their sense of this loss, directed, at a
meeting on February 23rd, that the following minute be entered on
the records of the corporation:

> "It is with sad hearts that we record the death of Mr.
> Edward B. Hatch, who died on February 18th, 1921,
> following a very brief illness. As a director of this
> Company since February, 1915, he ever showed his
> devoted interest in its affairs. He was sound in Christian
> character, kindly in nature, practical in business, and
> was regarded as a friend by all who knew him.
>
> The members of the Board of Directors, with a deep
> sense of their loss, desire to express to his wife their
> sincerest sympathy, and to record this minute upon the
> books of the Company, and the Secretary is hereby
> instructed to send a copy of it to Mrs. Hatch."

OUR "BOILER BOOK."

FOR more than forty years the designs of the Engineering Department of this Company have been accepted as standards throughout the United States and have been copied into the most
authoritative engineering handbooks and textbooks as well as the trade
catalogs of the best known boiler makers and dealers in materials
required for boiler construction.

Our designs and standards have been disseminated principally in
the form of blue-prints, of which more than 83,000 have been sent
out since the Engineering Department was established. As a rule,
however, each blue-print covers but one single phase of some particular subject and we have often thought that it would be worth
while to assemble some of the more commonly used data in the form
of a pamphlet. This idea has been more forcibly brought to our attention since the promulgation of the Boiler Code of the American
Society of Mechanical Engineers and its adoption by several states
and cities. Upon its publication, we adopted this Code as a standard
and all our drawings, tables, and other data in use today have been
designed in accordance with its provisions. In view of the many inquiries that we receive regarding the A. S. M. E. rules for boiler
design, it would seem that such data ought to be especially valuable to
those who are trying to follow such requirements.

With these ideas in mind we have published, "The Boiler Book" of The Hartford Steam Boiler Inspection and Insurance Company. This book, which has been gotten out in loose-leaf form so that it may later readily be expanded, is not intended as a treatise on boiler design but merely as a collection, in convenient form, of data which we have found valuable in our Engineering Department and which we hope will prove of equal value to those who may have occasion to use it. Among the subjects covered the following may be mentioned: Design of Riveted Joints; Tube Layouts; Allowable Pressures on Spherical Heads and Unstayed Furnaces; Design of Horizontal Tubular Boiler Settings with table of dimensions and number of brick required; tables relating to Heating Surface, Horse Power and Weights of Horizontal Return Tubular Boilers; Safety Valves; and other miscellaneous tables of interest. Since the subject matter deals only with the design of boilers the book is not one which would be of interest to the operating man. It is intended rather for the use of boiler-makers and designers, consulting engineers, and college professors.

We have a limited number of copies of The Boiler Book for distribution, at the price of $1.50 per copy, to those interested in it.

Pat was Persistent.

PAT was never happier than when in a heated argument. He lived to argue, he fattened on it, thrived on it.

At one time Pat had a job with Jones Brothers and Company as a fireman. Now Pat had never taken the Hartford Correspondence Course for Firemen but he thought he knew how to burn coal. No one could tell him any better methods than he knew and used. The Engineer did his best to show him how to use his head to save his back but it was no use. Pat used more time in arguing than he did in firing. So the Engineer decided that Pat must go. He knew that there would be trouble if he told Pat he was discharged so in order to save time he wrote Pat a letter to that effect.

Pat left, but a few days later the Engineer walked in to find Pat at work in the boiler room.

"Pat," he said, "did you get a letter from me?"

"Oi did, sor," said Pat.

"Well, Pat, what did that letter say?"

"Indade and Oi read yor letter on the insoide and on the ootsoide. On the insoide it said Oi was foired, but bejabers, on the ootsoide it said 'Rayturn to Jones Brothers and Company in foive days.'"

Summary of Inspectors' Work for 1920.

Number of visits of inspection made	207,641
Total number of boilers examined	393,900
Number inspected internally	173,034
Number tested by hydrostatic pressure	9,376
Number of boilers found to be uninsurable	1,139
Number of shop boilers inspected	14,469
Number of fly wheels inspected	39,924
Number of premises where pipe lines were inspected	10,423

SUMMARY OF DEFECTS DISCOVERED.

Nature of Defects.	Whole Number.	Danger-ous.
Cases of sediment or loose scale	35,236	1,938
Cases of adhering scale	47,064	1,799
Cases of grooving	2,136	294
Cases of internal corrosion	22,382	1,133
Cases of external corrosion	11,999	983
Cases of defective bracing	964	288
Cases of defective staybolting	3,058	715
Settings defective	10,065	892
Fractured plates and heads	3,738	544
Burned plates	4,046	547
Laminated plates	459	37
Cases of defective riveting	1,216	268
Cases of leakage around tubes	14,884	1,586
Cases of defective tubes or flues	21,804	6,531
Cases of leakage at seams	5,650	533
Water gauges defective	5,126	877
Blow-offs defective	5,346	1,505
Cases of low water	454	168
Safety-valves overloaded	1,212	332
Safety-valves defective	2,119	435
Pressure gauges defective	7,540	733
Boilers without pressure gauges	715	152
Miscellaneous defects	5,526	773
Total	212,739	23,063

GRAND TOTAL OF THE INSPECTORS' WORK FROM THE TIME THE COMPANY BEGAN BUSINESS, TO JANUARY 1, 1921.

Visits of inspection made	4,940,994
Whole number of inspections (both internal and external)	9,783,103
Complete internal inspections	3,832,669
Boilers tested by hyrostatic pressure	376,489
Total number of boilers condemned	28,978
Total number of defects discovered	5,492,424
Total number of dangerous defects discovered	603,683

FLYWHEEL EXPLOSIONS.
DURING 1920

No.	MONTH	DAY	NATURE OF ACCIDENT	Killed	Injured	CONCERN	BUSINESS	LOCATION
1	Jan.	17	Flywheel exploded		4	Kasa William Colliery	Coal Mining	Pottsville, Pa.
2	Jan.	21	Flywheel exploded			Goodyear Metallic Rub. Shoe Co.	Rubber Shoes	Naugatuck, Conn.
3	Feb.	5	Flywheel exploded		1	George Hunt	Wood Saw	Sacramento, Calif.
4	Feb.	23	Flywheel exploded			Pacific Lumber Co.	Saw Mill	Scotia, Calif.
			See Locomotive for April, 1920					
5	Feb.	29	Flywheel exploded on gasoline motor fire engine				Fire Dept.	Milwaukee, Wis.
6	March	12	Centrifugal separator exploded			City of Milwaukee	Gas Plants	Philadelphia, Pa.
7	March	23	Broken pulley			United Gas Improvement Co.	Cotton Mills	North Grosvenor Dale, Conn.
						Grosvenor Dale Co.		
8	May	5	Broken pulley			Hodge Boiler Works	Boiler Mfgs.	E. Boston, Mass.
9	June	1	Turbine overspeeded and burst			Phœnix Mills	Knitting Mill	Little Falls, N. Y.
10	June	23	Flywheel exploded			Arkansas Light & Power Co.	Power Station	Stuttgart, Ark.
11	July	11	Broken pulley			Ash Grove Lime & Portland Cement Co.	Lime & Cement	Chanute, Kans.
12	Aug.	4	Broken pulley			Jessup & Moore Paper Co.	Paper Mill	Wilmington, Del.
13	Aug.	6	Ensilage cutter exploded			Harry Klock	Farm	Harlowtown, Mont.
14	Aug.	7	Flywheel explosion		1	C. B. & Q. R. R.	Railroad Shop	Centerville, Ia.
15	Aug.	1	Flywheel exploded			Scott Paper Co.	Paper Mill	Chester, Pa.
16	Aug.	25	Pulley exploded			MacAndrews & Forbes Co.	Licorice	Camden, N. J.
17	Aug.	26	Flywheel exploded		1	West Va. Pulp & Paper Co.	Paper Mill	Mech'n'ville, N. Y.
18	Sept.	1	Pulley exploded			Jessap & Moore Paper Co.	Paper Mill	Wilmington, Del.
19	Sept.	15	Flywheel explosion			Herman Schmidt	Cotton Gin	Kingsbury, Tex.
20	Sept.	20	Ensilage cutter exploded			Cleve Walker	Farm	Alexandria, Ind.
21	Sept.	28	Flywheel exploded		1	Kentucky Utilities Co.	Power Station	Richmond, Ky.
			See page 163 of this issue of The Locomotive					
22	Oct.	19	Flywheel exploded		2	J. W. Wells Lumber Co.	Saw Mill	Menomonie, Wis.
23	Oct.	25	Flywheel exploded			Lancaster Cotton Oil Co.	Cotton Gin	Lancaster, S. C.
			See Jan. 1921 issue of The Locomotive					
24	Oct.	28	Pulley exploded			Richard Borden Mfg. Co.	Cotton Mill	Fall River, Mass.

FLYWHEEL EXPLOSIONS.

(Continued)

No.	MONTH	DAY	NATURE OF ACCIDENT	Killed	Injured	CONCERN	BUSINESS	LOCATION
25	Oct.	29	Flywheel exploded		1	Girardville Gas Works	Gas Plant	Shenandoah, Pa.
26	Nov	8	Flywheel exploded			Stuttgart Rice Mills Co.	Rice Mill	Stuttgart, Ark.
			See Jan. 1921 issue of The Locomotive					
27	Nov	19	Flywheel exploded		1	Charles Curry	Gasoline engine	Manchester, O.
28	Nov	20	Two flywheels exploded			Acme Road Machinery Co.	Road machinery	Frankfort, N. Y.
29	Nov.	21	Flywheel exploded			Federal Plate Glass Co.	Glass Works	Ottawa, Ill.
30	Dec.	16	Flywheel exploded			American Strawboard Co.	Paper Mill	Quincy, Ill.
31	Dec.	31	Flywheel exploded	1		Municipal Light Plant	Power Station	Hannibal, Mo.

BOILER EXPLOSIONS.

(INCLUDING FRACTURES AND RUPTURES OF PRESSURE VESSELS)

MONTH OF JULY, 1920

No.	DAY	NATURE OF ACCIDENT	Killed	Injured	CONCERN	BUSINESS	LOCATION
320	2	Boiler accident		1	Pennsylvania R. R.	Railroad	Atlantic City, N. J.
321	2	Boiler explosion			Main St. Bakery	Bakery	Sandoval, Ill.
322	2	Boiler explosion		1	P. Rioscette and P. Guluotte		Springfield, Ill.
323	4	Boiler ruptured			Ford Colleries Co.	Coal Mining	Bairdford, Pa.
324	5	Accident to blow off pipe			City of Caldwell	Lgt. & Wat. Plant	Caldwell, Kan.
325	6	Broken header			Westinghouse Air Brake Co.	Air Brake Mfgs.	Wilmerding, Pa.
326	7	Boiler explosion			Eugene Sharp farm	Threshing mach.	Milford, Del.
327	7	Four tubes pulled out			Grant Leather Corp'n	Tanning extracts	Kingsport, Tenn.
328	7	Tube rupture			Philip Dietz Coal Co.	Coal & Ice Dealers	Brooklyn, N. Y.
329	12	Accident to blow off pipe			Interstate Water Co.	Water supply	Danville, Ill.
330	12	Section heating boiler cracked			City of Springfield	School	Springfield, Mass.
331	13	Ruptured fire sheet			Windsor Ice Co.	Ice & Cold Storage	Windsor, Mo.

		Nature of accident	Owner	Plant	Location
333	13	Boiler explosion	Sharp farm	Threshing mach.	Houston, Del.
333	13	Nine headers cracked	Ehret Magnesia Mfg. Co.	Magnesia covering	Port Kennedy, Pa.
334	13	Furnace collapse	Orinoka Mills	Weaving & Dyeing	Philadelphia, Pa.
335	14	Valve burst	Belton Coal Mining Co.	Coal Mining	Drakesboro, Pa.
336	15	Section of heating boiler cracked	Georgia State Normal School	Normal School	Athens, Ga.
337	16	Boiler explosion	George Becker	Fertilizers	Watseon, Ohio
338	16	Autogenously welded boiler exploded		Sawmill	McMinville, Tenn.
339	17	Boiler exploded	Keating farm	Threshing mach.	Ponca City, Okla.
340	17	Boiler of locomotive exploded		Railroad	Kellogg, Minn.
341	19	Tube and header ruptured	American Steel & Wire Co.	Steel Plant	Waukegan, Ill.
342	19	Steampipe accident	Steamship "Aquitania"	Steamship	At sea.
343	20	Failure of boiler stop valve	Clinton-Copeland Co.	Candy Mfgrs.	Burlington, Iowa.
344	21	Boiler exploded		Rice Mill	Bogaloosa, La.
345	21	Boiler of locomotive exploded	Pittsburgh & Lake Erie R. R.	Railroad	Brownsville Jct. Pa
346	22	Tube rupture	Hull & Dillon Packing Co.	Meat packers	Pittsburg, Kan.
347	22	Tube rupture	Drayton Mills	Cotton Mill	Spartansburg, S.C.
348	22	Boiler explosion	Charles Luit farm	Threshing mach.	Keokuk, Iowa.
349	22	Failure of pipe fitting	City of Benton	Municipality	Benton, Ark.
350	24	Boiler of locomotive exploded	Georgia & Alabama R. R.	Railroad	Richland, Ga.
351	24	Tube rupture	Armour & Co.	Meat packers	Chicago, Ill.
352	26	Tube rupture	S. A. Uvalde & Gulf R. R.	Railroad	Pleasanton, Tex.
353	28	Tube rupture	Newcomb Hotel Co.	Hotel	Duluth, Minn.
354	28	Tube rupture	Aetna Portland Cement Co.	Cement Mfgs.	Fenton, Mich.
355	29	Heating boiler exploded	John Lang	Residence	Sioux City, Ia.
356	29	Tube rupture	Thos. Potter Sons & Co.	Linoleum Mfgs.	Philadelphia, Pa.
357	30	Boiler of locomotive exploded		Railroad	Greenfield, Mass.
358	30	Ruptured fire sheet	Mayfield Coal & Ice Co.	Coal & Ice Co.	Mayfield, Ky.
359	31	Heating boiler exploded	H. P. Crane	Residence	St. Charles, Ill.
360	31	Tube rupture	American Steel & Wire Co.	Steel Plant	Joliet, Ill.

MONTH OF AUGUST, 1920

		Nature of accident	Owner	Plant	Location
361	2	Boiler of locomotive exploded	Arabia Granite Co.	Granite Quary	Lithonia, Ga.
362	3	Tube rupture	Union Light & Power Co.	Power Station	June. City, Kan.
363	3	Section of heating boiler cracked	University of Pittsburgh	University	Pittsburgh, Pa.
364	3	Tube rupture	Aetna Portland Cement Co.	Cement Mfg.	Fenton, Mich.
365	4	Economizer explosion	Aluminum Ore Co.	Aluminum	E. St. Louis, Ill.
366	4	Tube rupture	Delaware River Steel Co.	Steel Plant	Chester, Pa.

MONTH OF AUGUST, 1920 — Continued.

No.	Day	NATURE OF ACCIDENT	Kind	Injured	CONCERN	BUSINESS	LOCATION
367	4	Ruptured fire sheet		3	Peterson Beverage Co.	Beverages	Grand Rds., Mich.
368	4	Furnace sheet cracked			McCleary, Wallin & Crouse	Carpet Mfgrs.	Amsterdam, N. Y.
369	5	Broken headers and ruptured tube			Crown Oil & Refining Co.	Oil refiners	Houston, Tex.
370	8	Boiler exploded		2	T. Carlots	Threshing mach.	Ponca City, Okla.
371	8	Two headers cracked			D. G. Dery	Power House	Santa Fe, N. M.
372	10	Two sections heating boiler cracked				Silk Mill	Allentown, Pa.
373	12	Boiler exploded			J. Chambers	Threshing mach.	Marion, O.
374	13	Three tubes pulled from drum			West Penn Power Co.	Power Station	Wheeling, W. Va.
375	13	Boiler explosion			W. J. Parrish	Brick Mfg	Richmond, Va.
376	14	Boiler explosion			Robt. Erdman	Threshing mach.	Aberdeen, S. Dak.
377	17	Boiler explosion	=	3	Golley & Finley	Machine Shop	Lima, Ohio.
378	18	Ruptured fire sheet			Eldon Ice & Fuel Co.	Ice Plant	Eldon, Mo.
379	18	Four headers broken			Glen Willow Ice Mfg. Co.	Ice Plant	Glen Willow, Pa.
380	19	Boiler of locomotive exploded	=	1	Canadian Western Lumber Co.	Lumber	Vancouver, B. C.
381	21	Tube failure	=	1	Middle West Utilities Co.	Power Station	Mounds, Ill.
382	21	Ruptured boiler			Harrison Electric Co.	Power Station	Harrison, Ark.
383	24	Fire sheet bulged and cracked			City of Fairmont	Lgt. & Wat. Wks.	Fairmont, Neb.
384	24	Superheater drum cracked			Public Service Corp. of N. J.	Power Station	Newark, N. J.
385	25	Boiler exploded	=	1	McAuliffe Farm	Threshing mach.	Woodstock, Ill.
386	25	Tube rupture			Consumers Power Co.	Power Station	Flint, Mich.
387	25	Boiler exploded		2	W. Harbold farm	Threshing mach.	Marshalltown, Ia.
388	26	Tube failure		3	Wells Building	Office Building	Milwaukee, Wis.
389	26	Ruptured fire sheet			Peoples Ice & Cold Storage	Ice & Cold Storage	Claremore, Okla.
390	26	Plate bulged and cracked			Susquehanna Collieries Co.	Coal Mining	Shamokin, Pa.
391	27	Tube failure	=	1	National Sewing Machine Co.	Sew. Mach. Mgrs.	Belvidere, Ill.
392	27	Tube pulled out of drum			Rosiclare Lead & Fluorspar Co.	Fluorspar	Rosiclare, Ill.
393	27	Cracked header			Cleveland & Erie Elec. Ry. Co.	Electric Railway	Elk Park, Pa.
394	29	Tube rupture			Providence Gas Co.	Gas Works	Providence, R. I.
395	30	Tube rupture			Woodward Iron Co.	Iron Works	Woodward, Ala.
396	31	Crown sheet failed		2	Whitney Engineering Co.	Eng. Supplies	Tacoma, Wash.

MONTH OF SEPTEMBER, 1920

No.	Day	Nature of accident	Owner	Use	Location
397	1	Boiler exploded	Del-Tex Syndicate	Drilling for oil	Powell, Texas
398	2	Tube rupture	Los Angeles Gas & Elec. Co.	Power Station	Los Angeles, Cal.
399	2	Head forced off	G. H. Prettyman	Canning	Milford, Del.
400	2	Boiler ruptured	Oscar Nelson	Gasoline plant	Brnchs Stg W. Va.
401	3	Boiler of locomotive exploded	Baltimore & Ohio R. R.	Railroad	Proctor, W. Va.
402	3	Boiler exploded		Drilling for oil	Portsfalls, W. Va.
403	3	Tube ruptured	Public Service Corp. of N. J.	Power Station	Perth Amboy, N. J.
404	5	Boiler ruptured	Texas Ice & Cold Storage Co.	Ice Plant	Dallas, Texas.
405	7	Boiler of locomotive exploded	Detroit, Toledo & Ironton R. R.	Railroad	Springfield, O.
406	7	Boiler of locomotive exploded	Chicago & Alton R. R.	Railroad	Shirley, Ill.
407	7	Tube ruptured	Tennessee Paper Mills	Paper Mill	Chattanooga, Tenn
408	8	Boiler explosion	Stoner "Empire City"	Steamship	On Lake Erie
409	9	Fire sheet bulged and cracked	"City of Madison"	Power Plant	Madison, Neb.
410	10	Ruptured patch	Vernon Parish Lumber Co.	Saw mill	Kurthwood, La.
411	10	Boiler of locomotive exploded	Delaware & Hudson R. R.	Railroad	Lake Placid, N. Y.
412	10	Boiler of locomotive exploded	C. R. I. & P. R. R.	Railroad	Falcon, Colo.
413	10	Boiler of locomotive exploded	Island Creek Coal Co.	Coal Mining	Holden, W. Va.
414	12	Tube failure		Boiler Shop	Chattanooga, Tenn.
415	12	Accident to blow off pipe	Forsch Packing Co.	Canning	Norma, N. J.
416	13	Boiler explosion	Antlers Lumber Co.	Lumber Mills	Antlers, Okla.
417	15	Heating boiler failure		Tailor	David City, Nebr
418	16	Crown sheet failure		Lumber Mill	Ferrwood, Idaho
419	18	Accident to steam pipe	Blackwell Lumber Co.	Coke Plant	Rices Landing, Pa.
420	21	Boiler of locomotive exploded	H. C. Frick Coke Co.	Railroad	Mateon, N. Y.
421	21	Two sections of heating boiler cracked	New York Central R. R.	School	Danbury, Conn.
422	22	Boiler explosion	Town of Danbury	Meat packer	Urbania, Va.
423	22	Boiler bagged and ruptured	D. P. Grmels	Ice Plant	Morganfield, Ky.
424	23	Tube rupture	Greenwell Brothers	Paper Mill	Wilmington, Del.
425	23	Boiler explosion	Jessup & Moore Paper Co.	Saw mill	Cameron, O.
426	23	Boiler bagged and ruptured		Coal Mining	Black Lick, Pa.
427	24	Tube failure	Graff Bros. Co.	Tin Plate Mfgrs	New Castle, Pa.
428	24	Boiler explosion	American Sheet & Tin Plate Co.	Threshing mach	Stevens, Pa.
429	24	Tube failure			Fairport, O.
430	25	Boiler exploded	Diamond Alkali Co.	Chemical Plant	Niagara Falls, N. Y.
431	26	Tube failure	Usco Chemical Co.	Power Station	Scottsbluff, Nebr.
432	27	Tube failure	Interm'n'a Lgt. & Power Co. Robert Smith Ale Brewing Co.	Brewers	Philadelphia, Pa.

The Hartford Steam Boiler Inspection and Insurance Company

ABSTRACT OF STATEMENT, JANUARY 1, 1921

Capital Stock, . . $2,000,000.00

ASSETS.

Cash in offices and banks	$366,891.88
Real Estate	90,000.00
Mortgage and collateral loans	1,543,250.00
Bonds and stocks	6,188,435.00
Premiums in course of collection	728,199.44
Interest accrued	116,654.78
Total assets	9,033,431.10

LIABILITIES.

Reserve for unearned premiums		$4,512,194.11
Reserve for losses		205,160.80
Reserve for taxes and other contingencies		388,958.85
Capital stock	$2,000,000.00	
Surplus over all liabilities	1,927,117.34	
Surplus to Policy-holders		**$3,927,117.34**
Total liabilities		$9,033,431.10

CHARLES S. BLAKE, President.
FRANCIS B. ALLEN, Vice-President, W. R. C. CORSON, Secretary.
L. F. MIDDLEBROOK, Assistant Secretary.
E. SIDNEY BERRY, Assistant Secretary.
S. F. JETER, Chief Engineer.
H. E. DART, Supt. Engineering Dept.
F. M. FITCH, Auditor.
J. J. GRAHAM, Supt. of Agencies.

BOARD OF DIRECTORS

ATWOOD COLLINS, President,
Security Trust Co., Hartford, Conn.

LUCIUS F. ROBINSON, Attorney,
Hartford, Conn.

JOHN O. ENDERS, President,
United States Bank, Hartford, Conn.

MORGAN B. BRAINARD,
Vice-Pres. and Treasurer, Ætna Life
Insurance Co., Hartford, Conn.

FRANCIS B. ALLEN, Vice-Pres., The
Hartford Steam Boiler Inspection and
Insurance Company.

CHARLES P. COOLEY, President,
Society for Savings, Hartford, Conn.

FRANCIS T. MAXWELL, President,
The Hockanum Mills Company, Rockville, Conn.

HORACE B. CHENEY, Cheney Brothers
Silk Manufacturers, South Manchester,
Conn.

D. NEWTON BARNEY, Treasurer, The
Hartford Electric Light Co., Hartford,
Conn.

DR. GEORGE C. F. WILLIAMS, President and Treasurer, The Capewell
Horse Nail Co, Hartford, Conn.

JOSEPH R. ENSIGN, President, The
Ensign-Bickford Co., Simsbury, Conn.

EDWARD MILLIGAN, President,
The Phœnix Insurance Co., Hartford,
Conn.

MORGAN G. BULKELEY, JR.,
Ass't Treas., Ætna Life Ins. Co.,
Hartford, Conn.

CHARLES S. BLAKE, President,
The Hartford Steam Boiler Inspection
and Insurance Co.

WM. R. C. CORSON, Secretary,
The Hartford Steam Boiler Inspection
and Insurance Company.

Incorporated 1866.

Charter Perpetual.

INSURES AGAINST LOSS FROM DAMAGE TO PROPERTY AND PERSONS, DUE TO BOILER OR FLYWHEEL EXPLOSIONS AND ENGINE BREAKAGE

Department.	Representatives.
ATLANTA, Ga., . . .	W. M. FRANCIS, Manager.
1103-1106 Atlanta Trust Bldg.	C. R. SUMMERS, Chief Inspector.
BALTIMORE, Md., . .	LAWFORD & McKIM, General Agents.
13-14-15 Abell Bldg. . .	JAMES G. REID, Chief Inspector.
BOSTON, Mass., ...	WARD I. CORNELL, Manager.
4 Liberty Sq., Cor. Water St.	CHARLES D. NOYES, Chief Inspector.
BRIDGEPORT, CT., . .	W. G. LINEBURGH & SON, General Agents.
404-405 City Savings Bank Bldg.	E. MASON PARRY, Chief Inspector.
CHICAGO, Ill., . . .	J. F. CRISWELL, Manager.
209 West Jackson B'v'l'd	P. M. MURRAY, Ass't Manager.
	J. P. MORRISON, Chief Inspector.
	J. T. COLEMAN, Ass't Chief Inspector.
	C. W. ZIMMER, Ass't Chief Inspector.
CINCINNATI, Ohio, . .	W. E. GLEASON, Manager.
First National Bank Bldg.	WALTER GERNER, Chief Inspector.
CLEVELAND, Ohio, . .	H. A. BAUMHART, Manager.
Leader Bldg. . . .	L. T. GREGG, Chief Inspector.
DENVER, Colo., . . .	J. H. CHESTNUTT,
918-920 Gas & Electric Bldg.	Manager and Chief Inspector.
HARTFORD, Conn., . .	F. H. KENYON, General Agent.
56 Prospect St. . .	E. MASON PARRY, Chief Inspector.
NEW ORLEANS, La., . .	R. T. BURWELL, Mgr. and Chief Inspector.
308 Canal Bank Bldg. . .	E. UNSWORTH, Ass't Chief Inspector.
NEW YORK, N. Y., . .	C. C. GARDINER, Manager.
100 William St. . . .	JOSEPH H. McNEILL, Chief Inspector.
	A. E. BONNETT, Ass't Chief Inspector.
PHILADELPHIA, Pa., . .	A. S. WICKHAM, Manager.
142 South Fourth St. .	WM. J. FARRAN, Consulting Engineer.
	S. B. ADAMS, Chief Inspector.
PITTSBURGH, Pa., . .	GEO. S. REYNOLDS, Manager.
1807-8-9-10 Arrott Bldg. .	J. A. SNYDER, Chief Inspector.
PORTLAND, Ore., . .	McCARGAR, BATES & LIVELY,
306 Yeon Bldg. . .	General Agents.
	C. B. PADDOCK, Chief Inspector.
SAN FRANCISCO, Cal., .	H. R. MANN & Co., General Agents.
339-341 Sansome St. .	J. B. WARNER, Chief Inspector.
ST. LOUIS, Mo., . .	C. D. ASHCROFT, Manager.
319 North Fourth St.	EUGENE WEBB, Chief Inspector.
TORONTO, Canada, . .	H. N. ROBERTS, President, The Boiler Inspection and Insurance Company of Canada.
Continental Life Bldg. .	

The Locomotive

HARTFORD STEAM BOILER INSPECTION AND INSURANCE CO.

DEVOTED TO POWER PLANT PROTECTION

PUBLISHED QUARTERLY

| Vol. XXXIII. | HARTFORD, CONN., JULY, 1921. | No. 7. |

BOILER EXPLOSION AT LUMBER, SOUTH CAROLINA.

SHOWING HOW THE BOILER RIPPED OPEN.

Another Long Seam Boiler Explosion.

THE long seam boiler again demonstrated its ability to kill people and cause enormous property damage when a boiler of this construction exploded at the mill of The McKeithan Lumber Company, Lumber, South Carolina, on April 26th.

The mill had been idle for some time and was about to resume operations, the boiler plant having been fired up for this purpose. Shortly before a force of two hundred men were to report for work, the boiler exploded with terrific violence and if the accident had occurred a little later the casualty list would have been a long one. As it was, two men were killed and the mill almost entirely destroyed, the damage being estimated at about $30,000. A view of the wreckage appears on the front cover of this issue of THE LOCOMOTIVE.

The boiler was built with two sheets — one upper and one lower — with two horizontal lap seams running the length of the shell. The explosion, as is usual in this type of construction, was the result of a lap seam crack. Whereas a boiler subject to this type of defect is always treacherous, the long seam gives additional opportunity for an extremely violent explosion because the boiler may rip open from head to head. This is clearly shown on the opposite page in the illustration of the wrecked boiler at the McKeithan mill. Evidence of the fatal lap seam crack can also be clearly seen.

A Recent Engine Breakdown.

PRODUCTION at the Raymondville, N. Y., plant of the Remington Paper and Power Company was seriously interfered with a short time ago by reason of an engine breakdown. Had the engine been an old one, the accident might not have caused great surprise, but since it was a comparatively new machine — having been in operation less than a year — the break came when least expected and the case is a good illustration of the hazard that exists in operating any and every engine, whether it be new or old.

The initial failure occurred at the cross-head, which broke into several pieces. When this happened the piston and rod were free of restraint and when the valve, which was still in operation, admitted steam to the crank end of the cylinder, the piston was driven with terrific force against the cylinder head. The blow ruptured almost every cylinder-head stud and drove off the head itself, which landed, as shown in the illustration, on the floor nearby. The piston did not leave the cylinder but was damaged to such an extent that a new one had to be installed.

An attendant, who was standing nearby, shut off the steam from the engine as soon as it was possible for him to do so but not before further damage had been done. The connect-

THE CYLINDER HEAD END OF THE WRECK.

ing-rod, driven by the crank and being somewhat free at the cross-head end, whipped about until it broke the crank-pin brasses and then fell into the crank pit. In this position it did not leave sufficient room for the crank-pin to pass by, with the result that the crank-shaft was

THE CRANK END OF THE WRECK.

lifted bodily from the inboard bearing when the crank-pin struck t h e connecting-rod. The parts came to rest at this point as illustrated herewith. It can be clearly seen how the bearing-cap was lifted f r o m its place, the cap bolts having been elongated to permit this move-ment.

To repair the en-gine, a new c r o s s-head, cylinder-head, piston and crank-pin were required and the mill was, of course, without the services of the engine while these new parts were being obtained and installed. The assistance of our inspector at the mill and of our branch office nearest the engine manu-facturer, however, served to expedite the repairs and, furthermore, the owners, being covered by both a Direct and a Use and Occupancy Engine Policy in The Hartford, were promptly indemnified.

The Value of Inspections.

NOT long ago one of our inspectors made a visit to a plant for an internal inspection of a horizontal return tubular boiler, and while in the furnace he noticed a brownish stain on the shell, well up on the side where the brick wall of the furnace and the shell join. To his trained eye this indicated a leak and he there-upon questioned the chief engineer as to whether any leakage had been noticed. The reply was in the negative but our inspector, nevertheless, was not satisfied. Climbing on top of the boiler he pro-

ceeded to learn, if possible, the source of the leak that had caused
that stain on the boiler shell.   The engineer called his attention to a
water pipe running across and somewhat above the boiler, mentioning that considerable condensation collected on this pipe and that
possibly there was some water reaching the boiler from this source.
This did not seem an altogether satisfactory explanation, however,
so the inspector ordered some of the asbestos covering of the boiler
removed at a point where he believed a leak might be.   The boiler
was then filled with water under pressure and it was probably some-

WATER ISSUING FROM A LAP SEAM CRACK.

what of a surprise to all present but the inspector to see water spray-
ing out through a crack in the boiler shell as shown in the photograph herewith.   The leakage was through a lap seam crack and it
is practically certain that this boiler would have caused a disastrous
explosion if it had remained in service.   The watchful eye of the inspector, however, prevented further use of the boiler and thereby
safeguarded life and property.

   The case just cited brings to our mind a somewhat similar one in
which the owner had repairs made and only by good fortune was a
serious accident averted.   The engineer of the plant, a laundry, noticed some steam issuing through the covering of the boiler and upon
investigation found that the steam came from a crack in the shell of

the boiler. The crack was of the lap seam type but the owner did not know its dangerous character and neither did the self-styled "boiler-maker" he called in for advice. This "expert" said, "That's all right. Go ahead and use her till the end of the week and then shut her down. I'll be over Sunday and weld her up." He apparently did not know that autogenous welding in a case of this kind is a most dangerous practice, or else he had no regard for the safety of the fifty girls at work directly above the boiler.

One of our inspectors heard of the case and made a visit to the place. As soon as he saw the nature of the leak he advised an immediate shutdown of the boiler and also told the owner that the contemplated repairs would not make the boiler any less dangerous to operate.

A few days later a hydrostatic test was applied to the boiler and the results were practically identical with the first case mentioned. The demonstration and the inspector's explanation of the serious nature of the defect were sufficient for the owner, and not only the leaking boiler but also its mate, both of them of the same age, were removed and new ones installed. Had this precaution not been taken there might very easily have been a repetition of the accident which occurred at the American Palace Laundry, Buffalo, N. Y., on November 3rd, 1906. In this disaster, the boiler, which was of the long seam type, failed as the result of a lap seam crack and four persons were killed. The property loss amounted to $12,000.

ENGINE ALIGNMENT

THE alignment of the component parts of an engine is an extremely important factor in determining its serviceable life. Too great care and attention cannot be given to the erecting and assembling of an engine, for undue wear and breakage from repeated stress on the parts will surely result if misalignment exists. It is well, also, when the opportunity is presented, to check up the alignment of an engine that has been in use for any length of time. Settling of foundations, structural weakness within the engine, and wear resulting from long continued service will bring about conditions that only a complete overhauling and readjustment can remedy.

The majority of engines are built with a foundation or sole-plate extending under and supporting the several castings. In such a case, lines representing the center lines of the cylinder and the cross-head may readily be marked on the sole-plate which can then be set level and in its proper relation to the machinery it is to drive. After this has been accomplished the different parts of the engine can be placed and aligned with each other. In the case of a new engine of this type, the alignment is taken care of by the machined joints of the separate castings so that the only point requiring particular care is the setting of the outboard bearing. Some of the smaller types of engines are entirely self-contained and are shipped as one assembled unit from the factory. Such engines, when new, require no attention other than setting in correct relation with the machinery to be driven and with the crank shaft and cylinder level. On the other hand, the wear that results from long continued service will frequently require that such an engine be readjusted and realigned.

There are some engines, however, that have been built without sole-plates, and, instead, have the different castings resting directly upon the concrete foundation. The alignment of such an engine requires exceptional care because there is no level, machined surface, such as the sole-plate, to insure alignment of the different parts. The cylinder casting, cross-head barrel, main frame and outboard bearing must be set individually and securely fixed in position as the work proceeds. In this work, flat iron wedges are used between the castings and the foundation so that a space of from ½" to 1" will exist for the bed of cement grout.

Possibly each type of engine might call for a special method of procedure as regards its alignment. We will assume, however, for the purpose of discussion, that we are dealing with the type of construction that has just been mentioned — that without a sole-plate — and that the engine is to be erected for belt drive to a line shaft. Such a case would cover practically all the points involved in the alignment of engines, and, from the discussion of the subject, particular steps may be selected or modified to suit particular cases or conditions.

The steps to be taken in this work will first be enumerated in the order in which they should be taken and a further discussion of them will then be given. The procedure which, in general, will apply to a new engine or an old one being re-erected is as follows:

(1) From the position of the line shaft obtain a reference line to which the center line of the cylinder is to be made parallel.

(2) Set the cylinder so that its center line is level and parallel to the reference line.

(3) Set the cross-head slide so that its center line will coincide with that of the cylinder.

(4) Set the casting which carries the inboard bearing so that the center line of this bearing will be at exact right angles to the center line of the cylinder.

(5) Set the outboard bearing to bring the center line of the shaft level and at right angles to the center line of the cylinder.

(6) Assemble the remaining parts of the engine, taking care to check the correct relationship or alignment of each part as the work proceeds.

In the case of a self contained engine it is a simple matter to place the crank shaft parallel to the line shaft by the usual procedure of aligning belt wheels and pulleys. In the case of the type of engine under consideration, however, whereas the crank shaft could easily be placed parallel to the line shaft, such procedure would involve difficulties when the cylinder and cross-head slide are to be adjusted to their proper relation with the crank shaft. It therefore becomes necessary to erect a reference line which shall be perpendicular to the line shafting and to which the center line of the cross-head slide will be made parallel.

To secure this reference line the line shafting must be placed in alignment from hanger to hanger. The n e x t step is to drop a plumb line from two points of the line shaft to the floor, as shown in Fig. 1 at AA, and thus get the line BB which will necessarily be parallel to the shaft. From a point C (the selection of which will later be made apparent) measure off the distance CD equal to three feet. Then with a board or steel tape, swing an arc with center at C and radius

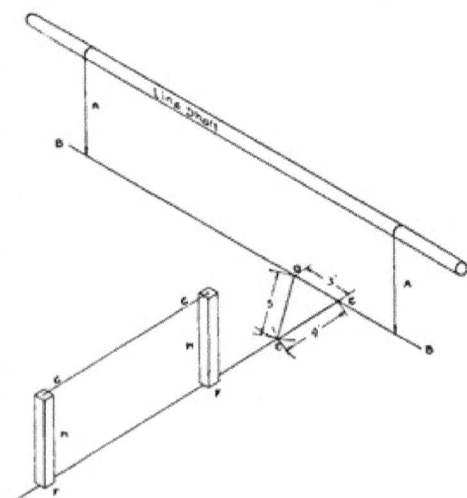

FIG. 1.

of four feet and another with center at D and radius of five feet.
The intersection of these two arcs will give the point E and the line
through CE will be perpendicular to the line BB because the sum of
the squares of the two short sides of a right angled triangle is equal
to the square of the hypotenuse or long side.

At the two points FF (the selection of which will also be evident
later on) erect two posts HH, plumb them vertical, and brace them to
remain fixed in that position. The tops of these posts should be at
about the same level as that of the cylinder. Now stretch a line,
GG, over the tops of the posts and, with the aid of a plumb bob or
spirit plumb, bring this line exactly over the line on the floor and
fasten it in this position. The line thus erected is, by construction,
perpendicular to the line shafting, although not intersecting it, and is
the reference line to which the axis of the cylinder is to be made
parallel.

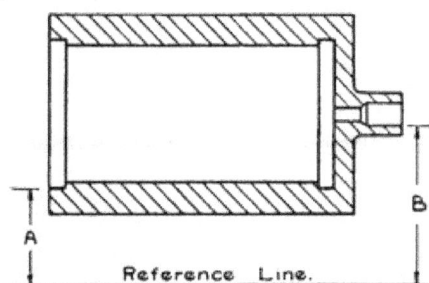

FIG. 2.

If the cylinder has not
already been placed in its
approximate position as
determined by the founda-
tion bolts, this part of the
work should now be at-
tended to and the cylinder
should then be set level and
parallel to the reference
line. A spirit level of
fairly good length placed
inside and on the bottom of
the bore will serve to show which way the cylinder must be shifted in
a vertical plane. If the cylinder has been in service for some time
and has become worn out of round it should be rebored. This work
should be entrusted only to competent men experienced in such work.

The locating of the cylinder in a horizontal plane so that it will
be parallel to the center line is a simple matter. Fig. 2 is a plan
view showing how this may be accomplished. The lengths A and B
are entirely arbitrary. What is of importance is the difference between A and B and this may readily be calculated from measurements of the cylinder.

The center line or axis of the cylinder is to be represented by a
stout linen thread or annealed copper wire stretched through the
cylinder, passing from the head end, through the piston rod stuffing

box, and continuing to a point well beyond the center line of the crank
shaft. This line must be located by careful measurement with reference to the bore of the cylinder at the head end and the bore of the
stuffing box at the crank end of the cylinder.

To locate and securely fix the line at the head end, a fixture

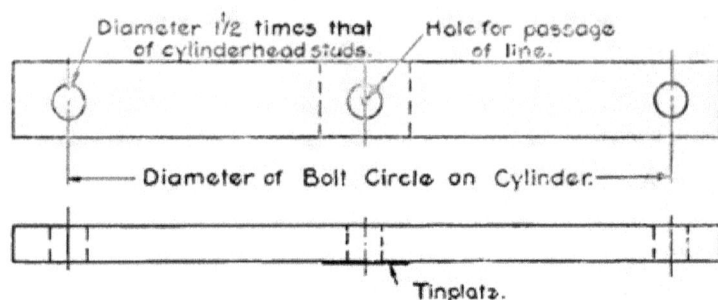

Fig. 3.

such as that shown in Fig. 3 may be used. For the crank end a wooden
plug similar to that shown in Fig. 4 should be made. This plug
should be turned and bored in the lathe so that the center of the
small hole passing through it and the center of the plug itself will
be identical. These two fixtures are installed on the cylinder as shown
in Fig. 5. One end of the line is tied around a short stick or rod,
as shown in Fig. 5, and is then passed from the head end of the
cylinder through the hole in the tin plate, which should be no larger
than to just permit passage of the line, through the hole in the plug
at the stuffing box and continued several feet beyond the center of

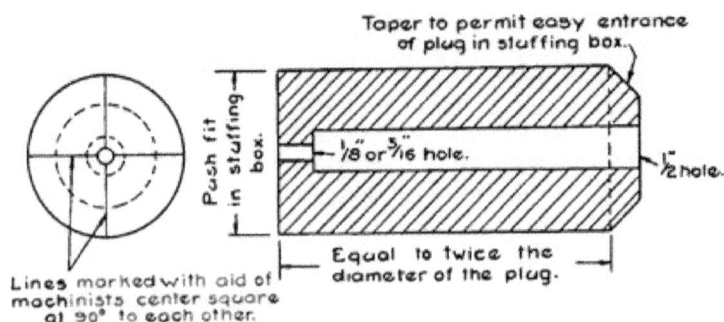

Fig. 4.

the crank shaft as indicated in Fig. 6, to a point on a standard or post
that has been erected in a thoroughly secure position for this purpose.
A convenient method for locating this end of the line — and the same

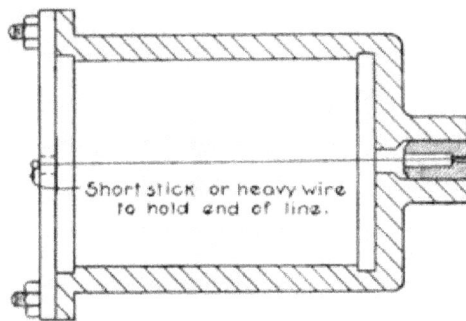

Fig. 5.

scheme may also be
used at the cylinder
end — is as illustrated
in Fig. 7 on page 204.
The notched piece of
sheet metal is clamped
to the post until its
correct position is ac-
curately determined
and it is then securely
fastened to the post.
The line is run
through this notch
and fastened on the
post at some convenient point. This method of providing two fixed
points through which the line may be run is adopted so that the line
may be removed or replaced when it is found necessary to do so.

The line must now be centered in the cylinder at both ends. At
the head end a centering rod or stick, with an ordinary pin driven in
each end, is used. By placing one end of this rod on the surface of
the cylinder bore at four points, 90° apart from each other, swinging the other end of the rod past or near the line and, at the same
time, shifting, if necessary, the position of the board bolted on the
cylinder, the line can be brought to the exact center of the cylinder

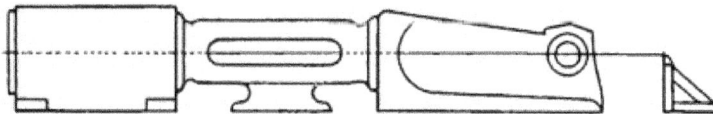

Fig. 6.

bore. The length of the test rod should be adjusted from time to
time until it will just touch the line when placed at the four equidis-
tant points. This is a very particular and nice part of the work and
must be carried out with extreme care. "About right" will not do
— it must be *exactly* right.

The method of centering the line by reference to the bore of the
cylinder should be used only where the cylinder is new or has been
rebored. In the case of an old engine which has not been moved

from its foundation but which is to be checked for alignment the centering of the line should be with reference to the counterbore of the cylinder. This is because wear of the cylinder may have occurred and an incorrect setting might thereby be made. Furthermore, this procedure provides an easy means of detecting any such wear.

If reboring is found necessary it must be remembered that the strength of the cylinder walls will be deceased so that a reduction of pressure may be necessary to provide a proper factor of safety.

When the head end of the line has been adjusted we may go to the crank end of the cylinder. In this case the position of the far end of the line must be shifted sideways or up and down until, by sighting across the reference cross lines on the plug in the stuffing box, the line is made to pass through the *exact* center. It is possible that this procedure may have changed the proper location of the line at the head end of the cylinder so it is advisable to go back to that point and check up. Before leaving this part of the work the line must be adjusted so that it is central at both ends simultaneously.

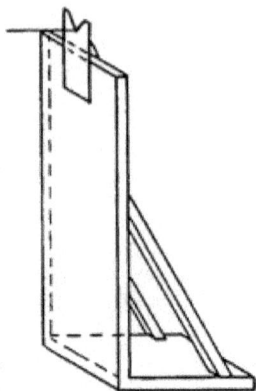

FIG. 7.

The notched pieces of sheet metal that have been used to locate the ends of the line should now be fixed in a secure manner in the position that has thus been determined for them. The line may then be removed from and replaced on these metal center points, whenever it is found necessary to do so, without the necessity of relocating each time it is replaced.

The alignment of the cross-head guide of the barrel type is a comparatively easy matter. The joint between the cross-head guide casting and the cylinder casting is usually made with a shoulder and counterbore so that the two parts are automatically centered when bolted together. All that is necessary with a new engine, therefore, is to set the guide level and bolt it tight to the cylinder. In the case of an old engine the surfaces of the slide itself should be tested for parallelism. This may readily be done with a pair of inside calipers, preferably of the micrometer type. If they are found parallel the same procedure may be followed as in the case of a new engine. If not parallel, the surfaces should be machined to bring them into the proper relation with each other and with the faces of the flanges at the ends of the casting. As in the case of the reboring of a cylinder, only a competent person should be entrusted with this work.

After the cross-head guide has been firmly fastened in position the main frame, carrying the inboard bearing, may be moved into place. Since the joint between the cross-head guide casting and the main frame is similar to the joint at the cylinder end of the guide, the alignment of the two parts in question is readily attained. Do not, however, attempt to pull the main frame into place by tightening up on this joint as such procedure would strain the cross-head slide casting if the two parts were not in exact alignment. The main frame casting should first be brought into perfect contact with the flange on the cross-head slide, secured in that position by the wedges and foundation bolts, and the flange joint then made tight.

The center line should now be placed in position and checked to see that it is correct, preparatory to aligning the shaft bearings.

In the type of engine under consideration, the outboard bearing is sometimes of the independent pedestal type with a sole-plate and means to raise and lower or to shift sideways the pedestal casting itself. In erecting this outboard bearing the sole-plate should be placed in its approximately correct position, with wedges between it and the foundation, and the pedestal casting then placed upon it with the adjusting wedges and screws set in about their middle position. With the bearing boxes or brasses in place we are ready to proceed with the alignment.

In this work the bearing caps are first removed and a piece of board is forced into the lower box of each bearing, as shown in Fig. 8. On this board a line is laid out, by careful measurement, to represent the center line of the bearing. A level line is then erected to intersect and to be at right angles to the center line of the cylinder. The bearings are then shifted so that their center lines lie exactly under and just touching the line of the shaft center. In doing this, any shifting of the outboard bearing that is necessary should be done, as far as possible, by moving or wedging the sole-plate and not by adjustments between the pedestal casting and the sole-plate. The distance between the two bearings must also be given careful attention.

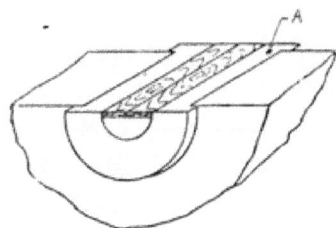

Fig. 8.

The erection of the shaft center line is probably the only point requiring further explanation at this point. Targets or posts, similar to the one used for the location of the crank end of the cylinder center line, should be used to carry the ends of the line in question. A large carpenter's square, known to be true, may be used to test the angle between the two lines. The 3-4-5 method, described earlier in this article, may also be applied in this case. A long spirit level may be used to level the line. In making these adjustments care should be taken to see that the lines are not touched so as to throw them out of position.

Instead of the board shown in Fig. 7 a circular piece, turned to the same diameter as the shaft with a small hole through the center and two cross lines marked upon it, may be used, one in each bearing. The idea is the same as that given in Fig. 4 for locating the cylinder center line through the stuffing box.

When all the foregoing steps in the alignment process have been carried out, the foundation bolts made tight and the alignment re-checked, the shaft center line may be removed and the engine grouted in place.

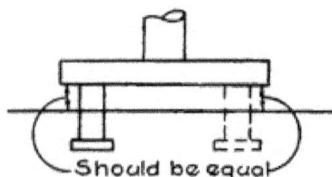

Should be equal

Fig. 9.

The shaft is then laid in its bearings and the boxes raised or lowered to bring it level, which condition may be indicated by a spirit level laid on the shaft. When the bearings have been correctly located in this respect they should be scraped to fit the shaft. This is a matter for the attention of an expert in such work.

Although every care may have been given to the work so far described, there may be some slight inaccuracies and it would be well to investigate a few points. The first of these may be the checking of the angle between the cylinder and the crank shaft center lines. This may easily be done by turning the shaft to the two positions shown in Fig. 9 and making the indicated measurements. If these distances are the same the shaft is correctly located. If not, the outboard bearing should be shifted in the proper direction to make them equal.

The cross-head, piston and piston rod should now be placed in position and adjustments made to align these parts. To do this the rod is connected to the cross-head while the parts are in the position shown in Fig. 9. The cross-head is then adjusted by means of the shoes at top and at bottom, until the rod is parallel to the slide. In the case of an old engine it is possible that the cross-head may need to be rebabbitted and refitted to the slide.

The condition produced by a crankpin that has been bent out of

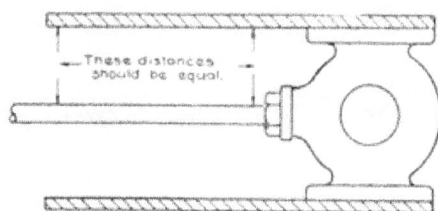

parallelism with the crank shaft is illustrated, in exaggerated form, in Fig. 11a. Such a condition will produce overheating and serious strains in the engine when it is running. The trueness of the crankpin in this respect may readily be observed by

Fig. 10.

turning the crankpin with the rod to one of the positions shown, measuring the distance A, and then swinging the parts to the other position and repeating the measurement. If the two distances, A, A agree, the crankpin is in correct alignment but if they do not agree some adjustments must be made either by scraping or by suitable machine work. Conditions vary so greatly that no exact procedure for correcting this type of fault can be prescribed. Since the crankpin may be bent in the direction shown in Fig. 11b the test should also be carried out in the positions shown there.

Another condition that may exist is that the center line of the crankpin brass may not be perpendicular to the center line of the connecting rod. To test for this, connect the rod to the crankpin, bring the

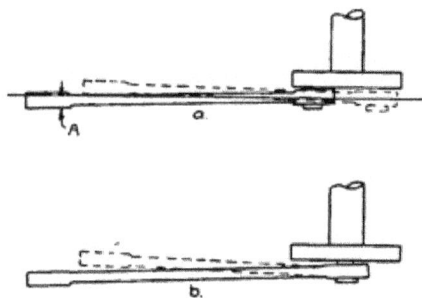

Fig. 11.

bearing brass snugly against the inner shoulder of the crankpin, and note the distance A as in Fig. 11a. Then disconnect the rod from the pin, turn it over half way and repeat the measurement. If the relation between the cylinder center line and the line on the rod is the same as in the first case the crankpin brass is in correct alignment.

The same test can be made on the wristpin brass in a similar way, measurements being made from the side of the connecting rod to the face of the crank disc. If in either case the bearings are found out of alignment, they should be scraped or machined to bring them to the correct condition.

It is possible that the wristpin may not be in correct alignment. This can be tested by connecting the rod snugly to the wristpin and then noting if the crankpin brass lies midway between the inside and outside collars or shoulders on the crankpin so that no possibility for binding exists on either side. This test should be repeated with the crank in several positions.

One more point should be checked before the engine is fully and finally assembled. This is in regard to the parallelism of the center lines of the connecting rod brasses. While the connecting rod is still coupled to the wrist pin, the crankpin bearing should be taken apart, the pin smeared with Prussian blue, and the bearing adjusted rather snugly on this pin. The engine is then turned over once and the bearing taken apart. If the bearing surface of the brass does not show an even bearing, as would be indicated by a fairly even deposit of blue over the whole surface, it should be scraped and retested until a good bearing is secured. The fitting and adjusting of the bearings requires judgment and only men of experience should be entrusted to do such work.

Before any of the bearings are finally bolted up, the bearing surfaces should be well covered with oil to insure lubrication until the usual provisions for oiling can come into full operation. When the engine is started and for some time afterwards, the bearings must be carefully watched for signs of overheating and if such trouble occurs the cause should be determined and remedied without delay. The same attention should also be given to any bearing that has been taken up for wear.

You, Mr. Watertender.

THE following was recently sent out by W. E. Thomson, steam-plant engineer of the Southern California Edison Co., to all watertenders of the company as an appeal to " sell them their jobs." It is well worth the attention of all boiler-plant employers.

Do you, Mr. Watertender, realize that you occupy an important position? You are largely responsible for the lives and safety of others as well as yourself. You have almost as much to do with the efficiency made by your shift as has the fireman. You are the one to see that the water is heated as hot as possible before it leaves the heater so that you gain by using all of the exhaust steam. Heating the water hot before it enters the boiler helps the boiler. It does not have to put so much heat into each pound of water, hence you increase the amount of steam the boiler can make. This increased capacity, especially over the peak load, enables your station to carry more kilowatts. Every eleven degrees you can heat the feed water by using exhaust steam means a saving of one per cent. in the amount of fuel oil used on your shift, and almost one per cent. gain in boiler capacity.

WATCH THE WATER LEVEL.

By watching the water level in the boiler, you keep the boiler from going dry, perhaps exploding. At one of our plants someone did not watch the water level. The water in a boiler was allowed to get low — way out of sight in the glass. A tube burst. Luckily, no one was injured, as the force of the explosion happened to be sideways into the other tubes. But the whole front bank of tubes had to be renewed. The brickwork had to be repaired. The total cost was over eleven hundred dollars, besides the loss from having the boiler out of service.

It is you, Mr. Watertender, who must watch to see that the water does not get too high in the glass. If it does get too high, the water will go over into the steam main. The temperature of the steam is lowered. More steam is required to carry the same load. More oil has to be used to make the steam. Hence your shift efficiency suffers. But this is not all. If enough water goes over, it is likely to wreck some machine. An instance of this happened not long ago. An exciter was wrecked. Fortunately, no one was hurt, but the repairs cost over five hundred dollars, and for some time, until the parts were received from the factory, the other exciters were overloaded and the plant in danger of shutting down any minute. This water, going over into the turbines, corrodes and scales the parts, causing a loss not only to your shift efficiency, but to everybody else's until the machine can be taken out and overhauled.

FEED THE WATER GRADUALLY.

It is you, Mr. Watertender, who can help your fireman and your shift efficiency by feeding the water into the boilers gradually, not have a valve half open one minute and closed the next, but set the valves so that the water goes into the boiler just as fast as the steamflow meter shows it is going out. It has been shown by tests that a swinging load will cause a loss of over 5 per cent., and that is just what you get when you feed the water into the boiler by spurts. You get the same action that a swinging load on the plant would cause. If you don't believe this, try it out. Take a reading on the flow meter, then open your feed valve wide and watch the flow meter — drops back, doesn't it? Now, close the feed valve and watch the flow meter — jumps right up, perhaps to a greater reading than you had at the start. Just the same action when you open the valve as if the station load would suddenly drop and the fireman had to cut back on his fires, only in this case the boiler stops steaming so fast, but you are using the same amount of fuel oil in the furnace. If the water is fed regularly, the flow-meter chart will not show any sudden swings and your shift efficiency will thus get the benefit.

CARRY WATER LEVEL AT HALF A GLASS.

Did you, Mr. Watertender, ever stop to thing that by carrying the water level constant at half a glass, your work is made easier and you are in a position to help the fireman out? A sudden demand comes for more steam. All right, you have a half-glass of water, a little extra, so you can shut the feed valve a minute — just long enough so that the boiler output increases because you are not putting cold water into it, but that minute gives the fireman a chance to get his steam pressure up. You can now open the feed a little — very gradually, remember, too fast will cause the steam to drop again — and slowly work your water levels up to half-glass again.

This little extra work on your part has kept the steam from getting away down so that both you and the fireman would have had to work for perhaps an hour to get it up. Now, suppose your water levels are back to half a glass again, and the fireman is carrying his steam high so as to get the best efficiency, a little load drops off, a boiler pops. All right, once more you can save. The boiler popping can stand a little more water; you open the feed valve. The popping stops almost immediately. The extra water you let in had to be heated. You have saved the steam that was going to waste through the pop valve. The fireman has now cut back on his fires a little so you can regulate the feed again until you have the half-glass of water showing in the gage.

BLOWING DOWN THE BOILERS.

It is you, Mr. Watertender, who is responsible for blowing down the boilers — for keeping the concentrate in the boilers below 200 so that the boilers will not prime. It is you who must keep the concentrate as near 200 as possible, so that heat will not be wasted by too much blowing down. Every time a boiler is blown down unnecessarily, it means a loss of approximately ten gallons of oil.

When a boiler is on stand-by, it is you, Mr. Watertender, who should report it in writing to the fireman if the boiler keeps filling up so you have to blow it down to keep the glass from getting full. It is you who should report it to the fireman if the stand-by boiler keeps losing water so you have to open the feed valve to keep the water in sight in the glass. In both these cases hot water is being wasted and you are in a position to catch these wastes before anybody else. A barrel of water wasted in either of these cases means a gallon of oil lost.

It is you, Mr. Watertender, who must blow down the water column and gage glass on each boiler at least once a shift so that you are sure the water level shown is correct. Otherwise, these lines may become clogged, the glass show water and still the boiler go dry, perhaps explode, kill someone and wreck the plant. —*from " Power."*

The class in French was reading an account of the war between France and Germany in 1870-71. The correct translation of a certain passage ran about as follows: " Thus the Germans had been able, without our having the slightest suspicion of it, to throw up gigantic works at a few thousand meters from our lines." Several students omitted the word " thousand " in translating which rendered the passage absurd enough, but one young lady improved on this by rendering: " a few millimeters from our lines." She knew, in a vague way, that " milli- " had *something* to do with thousands.

Similar mistakes may easily be avoided if one possesses a copy of The Metric System of Weights and Measures, published and for sale by The Hartford Steam Boiler Inspection and Insurance Co., Hartford, Conn., U. S. A. This publication is a valuable, indexed handbook of 196 pages of convenient size (3½" x 5¾") and substantially bound, containing a brief history of the Metric System, and *comparative tables* carefully calculated, giving the English or United States equivalents in all the units of measurement.

The price of the book, bound in sheepskin, is $1.25.

The Locomotive

DEVOTED TO POWER PLANT PROTECTION

PUBLISHED QUARTERLY

WM. D. HALSEY, EDITOR.

HARTFORD, JULY, 1921.

SINGLE COPIES *can be obtained free by calling at any of the company's agencies.*
Subscription price 50 cents per year when mailed from this office.
Recent bound volumes one dollar each. Earlier ones two dollars.
Reprinting matter from this paper is permitted if credited to
THE LOCOMOTIVE OF THE HARTFORD STEAM BOILER I. & I. CO.

IN the last issue of THE LOCOMOTIVE we published an article on foundations for engines and called particular attention to the importance of providing a well designed and well constructed foundation. In addition to a solid foundation there are a number of other conditions necessary for satisfactory operation and not the least of these is the correct alignment of the component parts of the engine. Faults in design, in material, and in workmanship present hazards in the operation of an engine that are many times mitigated by accurate alignment. And without alignment the best materials must sooner or later give way. In addition, even before an actual breakdown occurs, an amount of power difficult to estimate is wasted by the additional friction load produced.

In an endeavor to point out some of the principles of engine alignment we print, on another page, an article which we believe will be helpful. It would be difficult to cover every situation and discussion has therefore been limited to a specific case. The principles, however, are general and with a little thought may readily be applied to a problem of this nature.

" A sweet running engine " is the term applied by operating

engineers to an engine that runs smoothly and quietly. Perfect alignment of parts and adjustment of bearings is largely the secret of such operation.

The Flywheel is Still with Us.

One is prone to associate the flywheel with the old factory Corliss engine which drove by belt the machinery of a manufacturing establishment. The need of care in the operation of such a unit and the safety appliances necessary to safeguard the flywheel from disaster were generally appreciated. As the art has advanced in the last few years, the mind of the operator is turning to turbine units without flywheels, and the impression is gaining that the flywheel is not so important as it has been in the past. Occasionally, a flywheel is wrecked and we are brought face to face with the terrific amount of potential energy stored in one of these wheels and are forced to realize that the flywheel is still with us.

Where large turbine units are used with auxiliaries of the reciprocating type, there are a number of flywheels per unit, though of comparatively small size. It must be remembered that the energy stored in a wheel is not alone a function of the speed, but also depends upon the amount of metal that is in motion and the centrifugal force set up. Thus a machine with a comparatively small though heavy flywheel may have as much potential energy stored in it as a larger, lighter wheel.

Where, formerly, there was only the main flywheel to think of, with its accompanying safety device, there may be a number of flywheels serving one large generating unit, each one of them with its potentiality for disaster. Again, we must not neglect the flywheel effect of the rotor whether in the main unit or in the auxiliary machinery.

Failure of the flywheel on any one of the small units may cut a steam line or do damage that will cause a discontinuance of service to a much larger and more important load than formerly carried by one of the old-time factory engines. The introduction of centrifugal auxiliaries is reducing the number of flywheels in use, but at that there are few plants that do not contain a flywheel of some character. — *Editorial in " Power."*

The Veteran.

One of our contemporaries, *The Valve World*, published by The Crane Co., recently laid claim to the title of " the oldest corporation magazine in the United States that has been consistently and continuously a magazine (not a catalogue or a ' plant organ ') from the beginning." The claim was made upon the entry of *The Valve World* into its seventeenth year. In a spirit of indulgence we wrote the editor and called his attention to the fact that THE LOCOMOTIVE also came within his classification but that it dated from November 1867 — more than thirty-seven years before the first issue of *The Valve World*.

In a more recent issue of *The Valve World* the editor has replied to our letter and, in closing, says, " THE LOCOMOTIVE seems to be entitled to the palm as the ' oldest house organ ' — or corporation magazine — published in the United States, and we most heartily wish it continued success. It is an interesting publication and serves its own peculiar field as well and as faithfully as *The Valve World* endeavors to occupy its own chosen place."

The endeavors of *The Valve World* have been crowned with success and we hold it in high esteem. It is a most excellent publication, well written and well edited, and we feel honored by its expression of kind wishes. May it, too, continue to enjoy a life of service and success.

Luxurious Travel in the Olden Days.

THE frequency with which boilers blew up on the early Hudson River boats led to the use of what were known as " safety barges," and these, in their day, were considered the utmost luxury in travel, comparable to the private cars of the magnates of today. The barges were boats with main and upper decks and were almost as large as the steamers which towed them. The rabble rode on the steamers, inhaled the smells of the kitchen and the freight holds, endured the noise of the engines, and took the chances of explosions, while on the barges behind the elite traveled in luxurious state. Food was brought from the boat kitchen to the barge saloon over a swaying bridge between the vessels and was served with great aplomb under the direction of the barge captain, who was a noble figure in the setting.

The upper decks of the barges were canopied and decked with flowers, with promenades and easy chairs from which to view the scenery. At night the interiors were transformed into sleeping accommodations, much the same as a modern Pullman, except that they were more commodious. Not the least attractive feature of these barges, according to a chronicler of their excellence, was " an elegant bar, most sumptuously supplied with all that can be desired by the most fastidious and thirsty." — *Buffalo Courier*.

Boiler Explosions during 1920.

THE year 1920 produced the largest number of boiler accidents so far recorded. The total of accidents amounted to 652 which is greater by 102 than the largest previous year, 1909, when the number was 550. The number of persons killed and persons injured has, on the other hand, fallen below the average figure of recent years, 1920 showing a total of 137 killed and 262 injured.

The monthly figures for number of accidents, persons killed, persons injured, and total of killed and injured are given in the table below.

SUMMARY OF BOILER EXLOSIONS FOR 1920.

MONTH.	Number of Explosions.	Persons Killed.	Persons Injured.	Total of Killed and Injured.
January	86	13	27	40
February. . . .	66	12	20	32
March.	53	9	23	32
April	47	12	9	15
May . . . . :	36	5	11	16
June	31	8	21	29
July	41	16	22	8
August	36	5	29	34
September . . .	42	15	25	40
October	53	17	15	32
November . . .	87	18	22	40
December . . .	74	7	38	45
Totals . . .	652	137	262	399

BOILER EXPLOSIONS.

(INCLUDING FRACTURES AND RUPTURES OF PRESSURE VESSELS)

MONTH OF SEPTEMBER, 1920 (Continued)

No.	DAY	NATURE OF ACCIDENT	CONCERN	Killed	Injured	BUSINESS	LOCATION
433	27	Blow-off pipe failure	Boyle-Farrell Land Co.			Planing Mill	Farrell, Ark.
434	29	Collapse of crown sheet	Copeland-Inglis Shale Brick Co.			Brick Mfgrs.	Alton, Mo.
435	29	Boiler explosion				Threshing mach.	Marietta, Pa.
436	30	Section of heating boiler cracked	S. & H. A. Blumenthal			Apt. House	New York, N. Y.
437	30	Section of heating boiler cracked	University of Pittsburgh			University	Pittsburgh, Pa.
438	30	Section of heating boiler cracked	U. S. Government			Apt. House	New York, N. Y.

MONTH OF OCTOBER, 1920

No.	DAY	NATURE OF ACCIDENT	CONCERN	Killed	Injured	BUSINESS	LOCATION
439	1	Three headers ruptured	Clinton Woolen Mfg. Co.			Woolen Mill	Clinton, Mich.
440	2	Boiler of locomotive exploded	C. & N. W. R. R.			Railroad	Chadron, Nebr.
441	2	Five sections heating boiler cracked	T. H. Gaither & G. M. Balliere			Apt. House	Washington, D. C.
442	4	Rupture of hot water tank	R. S. Pollet			Residence	Worcester, Mass.
443	5	Ruptured boiler	Wellington Machine Co.			Mach. Mfgrs.	Wellington, O.
444	5	Steam pipe burst	Winchester Repeating Arms Co.		2	Arms & Ammunition	N. Haven, Conn.
445	6	Heating boiler exploded	People's Theatre Co.		3	Theatre	Greenville, Miss.
446	6	Section of heating boiler cracked	United Realty Co.			Apt. House	S. Francisco, Cal.
447	6	Section of heating boiler cracked	John Sheldon Estate			Office Building	Greenfield, Mass.
448	7	Mud drum ruptured	William Lea & Sons Co.			Flour Mill	Bloomington, Del.
449	10	Six headers cracked	Salmen Brick & Lumber Co.			Saw Mill	Slidell, La.
450	11	Section of heating boiler cracked	Wadsworth Village			School	Wadsworth, O
451	12	Fifteen headers broke	Kingston Coal Co.			Colliery	Kingston, Pa
452	14	Section of heating boiler cracked	The Citizen Co.			Publishers	Asheville, N. C.
453	15	Boiler exploded	Racey Gin Co.		5	Cotton Gin	Rowland, N. C.
454	15	Blow-off pipe ruptured	Terrell Cotton Oil Co.			Cotton Gin	Grand Salins, Tex
455	16	Boiler exploded	Wood-Bonnar Gin Co.		1	Cotton Gin	Honey Grove, Tex.
456	16	Five sections heating boiler cracked	H. P. Dygert			Theatre	E. Rochester, N. Y.

No.	Day	Description	Owner		Type	Location
457	17	Rupture of boiler shell	Paragon Plaster Co.		Plaster Works	Syracuse, N. Y.
458	17	Accident to steam pipe	Commonwealth Edison Co.		Public Utilities	Chicago, Ill.
459	19	Collapse of crown sheet	Mahoney Paving Co.		Road Building	Charlottesville, Va
460	19	Tubes pulled out of tube sheet	Pittsburgh Steel Products Co.		Tube Mill	Allenport, Pa.
461	20	Section of heating boiler cracked	J. R. Thompson		Restaurant	Buffalo, N. Y.
462	20	Blow-off pipe failure	Utah-Idaho Sugar Co.		Sugar Mfgrs.	Delta, Utah.
463	20	Boiler of locomotive **exploded**	Lehigh Valley R. R.	3	Railroad	Jersey City, N. J.
464	20	Tube failure	Kesler & Folse		Sugar Mfgrs.	Schotzville, La.
465	20	Section of heating boiler cracked	S. Liebovitz & Sons		Clothing Mfgrs.	Lawrence, N. Y.
466	21	Boiler of locomotive exploded	New York Central R. R.		Railroad	Batavia N. Y.
467	21	Boiler exploded	Tallahoma Lumber Co.		Lumber Mill	Laurel Miss.
468	21	Section of heating **boiler exploded**	George A. Fuller Co.	3	Hotel	Wilmington, N. C.
469	22	Valve body cracked	S. B. Wolf Shoe Co.		Shoe Mfg.	Cincinnati, O.
470	23	Heating boiler exploded	Morrow Hospital		Hospital	Seward, Nebr.
471	23	Boiler of locomotive exploded	C. R. I. & G. R. R.		Railroad	Fort Worth Tex.
472	23	Three sections heating boiler cracked	L. & M. Libman		Hotel	Hartford, Conn.
473	24	Nine sections **heating boiler cracked**	Calvary Church		Church	New York, N. Y.
474	25	Tube failure	Brewster & Co.		Carriage Mfgrs.	L. I. City, N. Y.
475	25	Two sections **heating boiler cracked**	Unw. Club Bldg. Ass'n		Club Building	Columbia, Mo.
476	25	Main stop valve **broken**	Brown Phelps Hosiery Co.		Hosiery Mfgrs.	Philadelphia, Pa.
477	25	Tube failure	Retail Grocers Ice Co.		Ice Plant	Little Rock. Ark
478	26	Tube pulled out	Bbit-Fibre Box Board **Mills**		Box Bd. Mfgrs.	Philadelphia, Pa.
479	27	Failure of fitting on blow-off pipe	Utah Idaho Sugar Co.		Sugar Mfgrs.	Delta, Utah.
480	27	Section of heating boiler cracked	United Realty Co.		Apt. Hotel	S. Francisco, Cal
481	28	Section of heating boiler cracked	Township No. 28, Range No. 5		School	Pontiac Ill
482	28	Three sections heating boiler **cracked**	F. Ruffolo.			Kansas City Mo
483	28	Tube failure	Snoqualmie Falls Lumber Co.		Lumber Mill	Snoqualmie Falls.
484	29	Heating boiler **exploded**	Oxford Apartments		Apt. House	Rochester, N. Y.
485	29	Heating boiler **exploded**	Jones & Holbeck			Chicago, Ill.
486	29	Steam pipe **failure**	Mary Graham Hotel Co.		Hotel	Atlantic City, N. J.
487	30	Mud drum **cracked**	Muessell Brewing Co.		Brewery	South Bend, Ind.
488	30	Four sections heating boiler **cracked**	Standish Realty Corporation		Theatre	New York N. Y.
489	30	Tube rupture	Public Service Corp. of N. J.		Power Station	Newark, N. Y.
490	30	Blow-off pipe failure	Lampasas Ice & Refrig. Co.		Ice Plant	Lampasas, Tex.
491	31	Explosion of diffuser tank	New England Power Co.	1	Paper Mill	Wilman Mills, Vt.

MONTH OF NOVEMBER, 1920

No.	DAY	Killed	Injured	NATURE OF ACCIDENT	CONCERN	BUSINESS	LOCATION
492	2		2	Boiler of locomotive exploded	B. & O. Railroad	Railroad	Akron, O.
493	2			Heating boiler exploded	New Capitol Hotel	Hotel	Little Rock, Ark.
494	2			Two sections heating boiler cracked	A. C. Johnson	Apt. House	Denver, Col.
495	3	1		Boiler exploded	Santa Fe Railroad	Hoisting Crane	Chicago, Ill.
496	3			Tube and six headers ruptured	Ford Motor Co.	Automobiles	H'land Pk. Mich.
497	4			Section of heating boiler cracked	Kane Borough School Board	School	Kane, Pa.
498	4		2	Heating boiler exploded	Delmonte Hotel	Hotel	St. Louis, Mo.
499	5			Two sections heating boiler cracked	B. & R. Gross	Apt. House	Hartford, Conn.
500	5			Boiler of locomotive exploded	Chicago North Western R. R.	Railroad	Iron Mt. Mich.
501	5			Nine sections heating boiler cracked	S. W. S. Amusement Co.	Theatre	Connellsville, Pa.
502	5			Tube ruptured	West Penn Power Co.	Power Plant	New York, N. Y.
503	5			Section of heating boiler cracked	Emigrant Industrial Savings Bk.	Bank	Greenville, Miss.
504	6			Heating boiler exploded	People's Theatre	Theatre	New York, N. Y.
505	6			Vulcanizer exploded	Spies Bros.	Auto Rep'r Shop	Cleveland, O.
506	6			Blow-off pipe ruptured	Knoxville Lumber & Mfg. Co.	Planing Mill	Knoxville, Tenn.
507	6			Tube ruptured	Utah-Idaho Sugar Co.	Sugar Mfrs.	Delta, Utah
508	6			Accident to blow-off valve	LeFlore Co. Gas & Elec. Co.	Light Plant	Poteau, Okla.
509	8			Section of heating boiler cracked	Pinehurst Hotel Co.	Hotel	Laurel, Miss.
510	8			Section of heating boiler cracked	D. of I. Home for Aged	Home	Newark, N. J.
511	9			Steam pipe burst	Crucible Steel Co.	Steel Mill	Pittsburgh, Pa.
512	9			Tube ruptured	Dubuque Elec. Co.	Power Plant	Dubuque, Ia.
513	9			Three sections heating boiler cracked	H. F. Bailey	Garage	Westfield, Mass.
514	10		1	Boiler of locomotive exploded		Railroad	Susquehanna, Pa.
515	10			Four sections heating boiler cracked	E. J. & H. W. Orr	Garage	Boston, Mass.
516	10			Section of heating boiler cracked	Twenty Morningside Ave. Corp.	Apt. House	New York, N. Y.
517	10			Accident to blow-off pipe	Blue Grass Laundry	Laundry	Dayton, Ky.
518	10			Five sections heating boiler cracked	Liberty Screw Co.	Factory	Worcester, Mass.
519	10			Section of heating boiler cracked	American Can Co.	Warehouse	New York, N. Y.
520	11			Section of heating boiler cracked	John Mayer	Apt. House	Kansas City, Mo.
521	11			Two sections heating boiler cracked	Pittsburg Machine Tool Co.	Machine Shop	Braddock, Pa.
522	11	2		Boiler exploded	C. P. Avery & Son	Drilling for Oil	Bingham, Ia.
523	11			Boiler shell ruptured	Alvah Nelson Lumber Co.	Planing Mill	Thomaston, Ga.
524	11			Three sections heating boiler cracked	Sam Harris	Apt. House	San Francisco, Cal.

No.		Accident	Owner	Occupancy	Location
525	12	Drier collapsed	Swift & Co.	Stock Yards	Chicago, Ill.
526	12	Section of heating boiler cracked	West Texas Nat'l Bank	Bank	Big Springs, Tex.
527	13	Tube ruptured	Barge "Pierce McLeouth"	Steam Barge	Alpena, Mich.
528	13	Seven sections heating boiler cracked	The Stanley Co. of America	Theatre	Philadelphia, Pa.
529	13	Accident to blow-off pipe	W. W. Carre Co.	Planing Mill	New Orleans, La.
530	14	Section of heating boiler cracked	Village of Atkinson	School	Atkinson, Ill.
531	14	Section of heating boiler cracked	I. T. S. Rubber Co.	Rubber Works	Elyria, O.
532	15	Boiler bulged and ruptured	Sherman Hotel Bldg. Corp'n	Somerset Hotel	Chicago, Ill.
533	15	Section of heating boiler cracked	L. Kauser	Apt. House	Hugoton, Tex.
534	15	Section of heating boiler cracked	Borough of Naugatuck	School	Union City, Conn.
535	15	Accident to blow-off pipe	St. Mary's Hospital	Hospital	Evansville, Ind.
536	15	Two sections heating boiler cracked	Borough of Indiana	School	Indiana, Pa.
537	15	Three sections heating boiler cracked	Louis Rutrup	Furniture Store	Hartford, Conn.
538	15	Tube ruptured	Warner-Klipston Chemical Co.	Chemical Works	Charleston, W.Va.
539	16	Two sections heating boiler cracked	Mobile Y. M. C. A.	Y. M. C. A.	Mobile, Al.
540	16	Section of heating boiler cracked	Magazner Baking Corp'n	Bakers	Springfield, Mass.
541	16	Two sections heating boiler cracked	J. R. Thompson	Restaurant	Chicago, Ill.
542	16	Nine tubes pulled out of drum	Delta Light & Traction Co.	Power Plant	Greenville, Miss.
543	16	Section of heating boiler cracked	Highland Hospital	Hospital	Asheville, N.C.
544	17	Accident to steam pipe	Hatfield & Penfield Steel Co.	Steel Foundry	Willoughby, O.
545	17	Four sections heating boiler cracked	Beaumont Floral Co.	Greenhouse	Beaumont, Tex.
546	18	Boiler exploded	Grant & Mays	Drilling for oil	Texarkana, O.
547	18	Tube ruptured	Taylor Wharton Iron & Steel Co.	Steel Mill	High Bridge, N.J.
548	19	Tube ruptured	Solvay Process Co.	Chemical Plant	Detroit, Mich.
549	19	Heating boiler exploded	Chilton Apartments	Apt. House	Richmond, Va.
550	19	Range boiler exploded	Augustus Stern	Residence	New York, N.Y.
551	19	Seven sections heating boiler cracked	L. & M. Libman	Residence	Hartford, Conn.
552	20	Section of heating boiler cracked	Wm. H. Trowbridge	Office Bldg.	Stafford, Mass.
553	21	Heating boiler exploded	Fred Wood	Residence	Saffron, Mass.
554	21	Heating boiler exploded	Nelson Johnson	Apt. House	Kansas City, Mo.
555	21	Header of heating boiler cracked	Wells Building Co.	Hotel	Boston, Tex.
556	23	Heating boiler exploded	Kyle Building Co.	Garage	Orchester, Mass.
557	23	Three sections heating boiler cracked	J. F. Rolfe	Office Bldg.	Salem, Mass.
558	24	Boiler exploded	Stoughton Marketing Co.	Market	Stoughton, Wis.
559	24	Two tubes ruptured	The National Tube Co.	Tube Mill	Lorraine, O.
560	24	Section of heating boiler cracked	Ashland Realty Co.	Net House	Buffalo, N.Y.
561	25	Boiler exploded	Oliver Oil Co.	Drilling for oil	Shelton, O.
562	25	Tube ruptured	Michigan Alkali Co.	Chemical Plant	Wyandotte, Mich.
563	25	Section of heating boiler cracked	James E. Bolin	Apartment House	Hartford, Conn.

MONTH OF NOVEMBER, 1920 — Continued.

No.	DAY	NATURE OF ACCIDENT	Killed	Injured	CONCERN	BUSINESS	LOCATION
564	26	Hot-water boiler exploded		2	Ida Weissbrott	Residence	New York, N. Y.
565	26	Boiler exploded,	1	3	Dunwoody Milling Co.	Flour Mill	Dunwoody, Ga.
566	26	Tube ruptured			Pittsburgh Plate-Glass Co.	Glass Works	Ford City, Pa.
567	27	Boiler of locomotive exploded		3	Seaboard Air Line Railroad	Railroad	W. Jacksville, Fla.
568	27	Tube ruptured	1		Barnsdall Zinc Co.	Zinc Mill	Waco, Mo.
569	27	Section heating boiler cracked			St. Andrews Church	Church	Buffalo, N. Y.
570	27	Accident to steam pipe			Peet Bros. Mfg. Co.	Soap Works	Kansas City, Mo.
571	27	Two sections heating boiler cracked			Sol Bettin	Apt. House	Houston, Tex.
572	28	Manifold and three sections of heating boiler cracked					
573	29	Two sections heating boiler cracked			City of Greenville	Hospital	Greenville, S. C.
574	29	Boiler exploded		2	Kingsley School	School	W. Waterloo, Ia.
575	29	Mud drum rupture				Donkey Engine	Galveston, Tex.
576	29	Section of heating boiler cracked			N. S. Coal & Coke Co.	Coal Mining	Gary, W. Va.
577	30	Autogenous weld failed			Fifteenth St. Amusement Co.	Theatre	Brooklyn, N. Y.
578	30	Section of heating boiler cracked			Holt Hotel Co.	Hotel	Lafayette, Ind.
					School District of Pittsburgh	School	Pittsburgh, Pa.

MONTH OF DECEMBER, 1920

No.	DAY	NATURE OF ACCIDENT	Killed	Injured	CONCERN	BUSINESS	LOCATION
579	1	Boiler exploded	1		Gustave Liljeroot Farm	Threshing Mach.	Galva, Ill.
580	1	Rupture to fitting on boiler feed line			Utah-Idaho Sugar Co.	Sugar Mfrs.	Delta, Utah
581	1	Rupture to fitting on blow-off pipe			Lawrence Hotel Co.	Hotel	Erie, Pa.
582	1	Accident to steam pipe			City of Monroe	Power Station	Monroe, La.
583	2	Tube ruptured		2	Standard Steel Works	Steel Plant	Lewistown, Pa.
584	2	Boiler plate bulged and ruptured			Aberdeen Gas Co.	Gas Plant	Aberdeen, S. D.
585	3	Tube ruptured			American Steel Wire Co.	Steel Plant	Anderson, Ind.
586	3	Boiler plate bulged and ruptured			Iowa Methodist Hospital	Hospital	Des Moines, Ia.
587	3	Section of heating boiler cracked			F. B. Jennings	Store Building	Fall River, Mass.
588	4	Section of heating boiler cracked			Rhinelander Garage	Garage	New York, N. Y.
589	4	Tube ruptured			Los Angeles Gas & Elec. Corp'n	Power Plant	Los Angeles, Cal.
590	6	Tube ruptured		2	Consumers Power Co.	Power Plant	Gd. Rapids, Mich.

No.	Day	Accident	Owner	Business	Location
591	6	Two sections heating boiler cracked	Sonkea-Galamla Iron & Metal Co	Iron & Metal Co	Kansas City, Kan
592	6	Stop-valve ruptured	Southport Mill, Ltd	Oil Refining	New Orleans, La
593	6	Accident to steam-pipe fitting	C. H. Guenther & Son	Flour Mill	San Antonio, Tex
594	7	Tube ruptured	1 F. R. Ward School	School	DuQuoin, Ill
595	7	Accident to steam pipe	2 F. D., D. M. & S. Interurban Co.	Power Plant	Beaue, Ia.
596	7	Boiler ruptured	Vernon Parish Lumber Co.	Lumber	Kurthwood, La.
597	7	Two sections heating boiler cracked	J. H. Levy & Sons, Inc.	Apt. House	New York, N.Y.
598	7	Tube ruptured and headers cracked	Warren Manufacturing Co.	Paper	Wren Mills, N.Y.
599	8	Header ruptured	Pittsburgh Plate Glass Co	Glass Mfrs	Ford City, Pa
600	10	Boiler exploded	C. N. Lindes	Laundry	Menasha, Wis.
601	10	Stop-valve ruptured	Utah-Idaho Sugar Co.	Sugar Mfrs	Payson, Utah
602	11	Section of heating boiler cracked	G. F. Heublein	Garage	Hartford, Conn.
603	11	Eight sections heating boiler cracked	G. W. Welch	Restaurant	Omaha, Nebr
604	11	Heating boiler exploded	Louis Fortier	Residence	Jenkintown, Pa
605	12	Five headers cracked	Freeport & Mexican Fuel Oil Op'n	Oil Refining	Arabi, La
606	12	Stop-valve ruptured	Rand Avery Supply Co.	Printers	Boston, Mass.
607	13	Three hot-water tanks exploded		Apt. House	Portland, Me.
608	13	Section of heating boiler cracked	City of Lincoln	School	Lincoln, Nebr
609	14	Boiler exploded	Waldon Oil Co.	Drilling for oil	Beaumont, Tex
610	14	Boiler exploded	Fang Farm	Saw Mill	Cox's Mills, W.Va
611	14	Section of heating boiler cracked	Lemon Thompson Realty Corp'n	Office Bldg	Glens Falls, N.Y.
612	14	Two sections heating boiler cracked	E. J. & F. W. Orr	Garage	Newtonville, Mass.
613	15	Section of heating boiler cracked	Stapley Co. of America	Theatre	Philadelphia, Pa.
614	15	Section of heating boiler cracked	Colonial Scars	Apt. House	Bridgeport, Conn.
615	15	Section of heating boiler cracked	Borough of Indiana	School	Indiana, Pa
616	15	Tube ruptured			Gd. Rapids, Mich
617	15	Two sections heating boiler cracked	L. Sarbinski	Apt. House	Akron, O.
618	15	Section of heating boiler cracked	The Bland Hotel	Hotel	Raleigh, N.C
619	17	Fire sheet ruptured	Breese Trenton Coal Co	Coal Mining	Beckemeyer, Ill.
620	17	Tube ruptured	Sumter Lumber Co.	Lumber	Delta Mills, Miss
621	18	Rupture of fitting on blow-off pipe	Utah-Idaho Sugar Co.	Sugar Mfrs	Delta, Utah
622	19	Boiler exploded		Apt. House	Elizabeth City, N.C
623	19	Two sections heating boiler cracked	Bellman Brook Bleachery Co.	Bleaching Plant	Fairview, N.H
624	20	Accident to steam pipe	Hughes Realty & Investment Co.	Office Bldg	Richmond, Mo
625	21	Tube blew out	Los Angeles Gas & Elec. Corp'n	Power Plant	Los Angeles, Cal
626	22	Section of heating boiler cracked	Borough of Indiana	School	Indiana, Pa
627	22	Boiler exploded. Resulting fire destroyed building	Feitschan's Junior High School	School	Springfield, Ill.

The Hartford Steam Boiler Inspection and Insurance Company

ABSTRACT OF STATEMENT, JANUARY 1, 1921

Capital Stock, . . $2,000,000.00

ASSETS.

Cash in offices and banks	$366,891.88
Real Estate	90,000.00
Mortgage and collateral loans	1,543,250.00
Bonds and stocks	6,188,435.00
Premiums in course of collection	728,199.44
Interest accrued	116,654.78
Total assets	9,033,431.10

LIABILITIES.

Reserve for unearned premiums		$4,512,194.11
Reserve for losses		205,160.80
Reserve for taxes and other contingencies . . .		388,958.85
Capital stock	$2,000,000.00	
Surplus over all liabilities	1,927,117.34	

Surplus to Policy-holders . , . . $3,927,117.34

Total liabilities	$9,033,431.10

CHARLES S. BLAKE, President.
FRANCIS B. ALLEN, Vice-President,   W. R. C. CORSON, Secretary.
L. F. MIDDLEBROOK, Assistant Secretary.
E. SIDNEY BERRY, Assistant Secretary.
S. F. JETER, Chief Engineer.
H. E. DART, Supt. Engineering Dept.
F. M. FITCH, Auditor.
J. J. GRAHAM, Supt. of Agencies.

BOARD OF DIRECTORS

ATWOOD COLLINS, President,
  Security Trust Co., Hartford, Conn.
LUCIUS F. ROBINSON, Attorney,
  Hartford, Conn.
JOHN O. ENDERS, President,
  United States Bank, Hartford, Conn.
MORGAN B. BRAINARD,
  Vice-Pres. and Treasurer, Ætna Life
  Insurance Co., Hartford, Conn.
FRANCIS B. ALLEN, Vice-Pres., The
  Hartford Steam Boiler Inspection and
  Insurance Company.
CHARLES P. COOLEY, President,
  Society for Savings, Hartford, Conn.
FRANCIS T. MAXWELL, President,
  The Hockanum Mills Company, Rock-
  ville, Conn.
HORACE B. CHENEY, Cheney Brothers
  Silk Manufacturers, South Manchester,
  Conn.

D. NEWTON BARNEY, Treasurer, The
  Hartford Electric Light Co., Hartford,
  Conn.
DR. GEORGE C. F. WILLIAMS, Presi-
  dent and Treasurer, The Capewell
  Horse Nail Co., Hartford, Conn.
JOSEPH R. ENSIGN, President, The
  Ensign-Bickford Co., Simsbury, Conn.
EDWARD MILLIGAN, President,
  The Phœnix Insurance Co., Hartford,
  Conn.
MORGAN G. BULKELEY, JR.,
  Ass't Treas., Ætna Life Ins. Co.,
  Hartford, Conn.
CHARLES S. BLAKE, President,
  The Hartford Steam Boiler Inspection
  and Insurance Co.
WM. R. C. CORSON, Secretary,
  The Hartford Steam Boiler Inspection
  and Insurance Company.

Incorporated 1866.

Charter Perpetual.

INSURES AGAINST LOSS FROM DAMAGE TO PROPERTY AND PERSONS, DUE TO BOILER OR FLYWHEEL EXPLOSIONS AND ENGINE BREAKAGE

Department.	Representatives.
ATLANTA, Ga.,	W. M. Francis, Manager.
1103-1106 Atlanta Trust Bldg.	C. R. Summers, Chief Inspector.
BALTIMORE, Md.,	Lawford & McKim, General Agents.
13-14-15 Abell Bldg.	James G. Reid, Chief Inspector.
BOSTON, Mass.,	Ward I. Cornell, Manager.
4 Liberty Sq., Cor. Water St.	Charles D. Noyes, Chief Inspector.
BRIDGEPORT, CT.,	W. G. Lineburgh & Son, General Agents
404-405 City Savings Bank Bldg.	E. Mason Parry, Chief Inspector.
CHICAGO, Ill.,	J. F. Criswell, Manager.
209 West Jackson B'v'l'd	P. M. Murray, Ass't Manager.
	J. P. Morrison, Chief Inspector.
	J. T. Coleman, Ass't Chief Inspector.
	C. W. Zimmer, Ass't Chief Inspector.
CINCINNATI, Ohio,	W. E. Gleason, Manager.
First National Bank Bldg.	Walter Gerner, Chief Inspector.
CLEVELAND, Ohio,	H. A. Baumhart, Manager.
Leader Bldg.	L. T. Gregg, Chief Inspector.
DENVER, Colo.,	J. H. Chesnutt,
916-918 Gas & Electric Bldg.	Manager and Chief Inspector.
HARTFORD, Conn.,	F. H. Kenyon, General Agent.
56 Prospect St.	E. Mason Parry, Chief Inspector.
NEW ORLEANS, La.,	R. T. Burwell, Mgr. and Chief Inspector.
308 Canal Bank Bldg.	E. Unsworth, Ass't Chief Inspector.
NEW YORK, N. Y.,	C. C. Gardiner, Manager.
100 William St.	Joseph H. McNeill, Chief Inspector.
	A. E. Bonnett, Ass't Chief Inspector.
PHILADELPHIA, Pa.,	A. S. Wickham, Manager.
142 South Fourth St.	Wm. J. Farran, Consulting Engineer.
	S. B. Adams, Chief Inspector.
PITTSBURGH, Pa.,	Geo. S. Reynolds, Manager.
1807-8-9-10 Arrott Bldg.	J. A. Snyder, Chief Inspector.
PORTLAND, Ore.,	McCargar, Bates & Lively,
306 Yeon Bldg.	General Agents.
	C. B. Paddock, Chief Inspector.
SAN FRANCISCO, Cal.,	H. R. Mann & Co., General Agents.
339-341 Sansome St.	J. B. Warner, Chief Inspector.
ST. LOUIS, Mo.,	C. D. Ashcroft, Manager.
319 North Fourth St.	Eugene Webb, Chief Inspector.
TORONTO, Canada,	H. N. Roberts, President, The Boiler Inspection and Insurance Company of Canada.
Continental Life Bldg.	

It's all In the Firing!

Wasteful Fuel Consumption! High Steam Cost!

Correspondence Course

For FIREMEN

A CONSTRUCTIVE SERVICE

extended to anyone interested

in BOILER ECONOMY and SAFETY

Write to day for details to the HOME OFFICE of

THE HARTFORD STEAM BOILER
INSPECTION and INSURANCE CO.

HARTFORD   CONNECTICUT

The Locomotive

Devoted to Power Plant Protection

Published Quarterly

| Vol. XXXIII. | HARTFORD, CONN., OCTOBER, 1921. | No. 8 |

HEAD OF EXPLODED LAUNDRY MANGLE.

Explosion of a Laundry Mangle.

The guests of the Hotel Pemberton and the residents of Hull, Mass., had an element of terror injected into their peaceful summer life on August 11th by the explosion of a mangle in the laundry of the hotel. A newspaper account in *The Boston Post* for August 12th stated that the explosion wrecked the laundry, tore an eight foot hole in the hotel wall as great pieces of the mangle, weighing hundreds of pounds, were hurled through it, and scattered debris all about the New Haven Railroad station and the Nantasket Beach Steamboat pier. Ten persons were injured in the accident, eight of them being

SHOWING FLIGHT OF THE MANGLE HEAD.

hotel employees and the other two being guests who were sitting on the porch just above the laundry at the time of the accident. The property damage amounted to about $4000.

It was said that a few days prior to the explosion steam was seen escaping through a crack in the cast iron shell close to one of the heads of the mangle. To stop this leak, autogenous welding was resorted to and while this method of repair temporarily stopped the escape of steam it would appear that it materially weakened the shell. Examination of the broken parts showed that the casting had been cracked for some time at the base of the flange, as shown in Fig. 1, this defect extending around the circumference meeting the welded portion at both ends, and varying in depth from one eighth to three eighths of an inch from the inside surface of the shell. The extent of this crack at the point where the leak developed could not be determined because the welding had destroyed all traces other than those mentioned above. The welded portion extended for about twenty-five inches around the circumference and was about five inches wide but its holding power was probably slight since a large portion of it appeared defective. At the break, the larger part of the weld had the appearance of slag which indicated that the metal had not fused at all, excepting on the outside surface. While no doubt the crack which already existed in the mangle cylinder would, in time, have caused a failure, it is, on the other hand, very likely that the welding hastened the end.

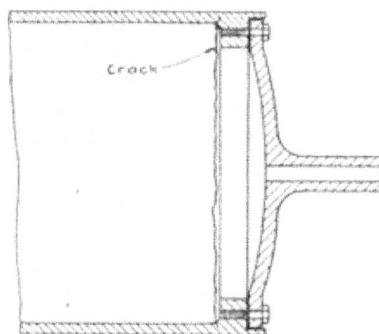

Fig. 1.

The head which broke away weighed approximately 750 pounds and was blown through the brick wall of the laundry and across the street, a distance of about sixty feet, coming to rest near the ticket office in the railroad station. The flight which it took is illustrated in the picture on the opposite page, the head lying at the spot marked 2 and the hole in the hotel wall, boarded up when the photograph was taken, at the spot marked 1. The head itself is shown on the front cover of this issue of THE LOCOMOTIVE. The cylinder and the other head were driven in the opposite direction through the basement for a distance of about fifty feet.

It is evident from this that a laundry mangle may become a rather powerful engine of destruction and therefore should be subjected to rigid inspection and given the greatest care in its operation.

More About the Value of Inspections.

TWO instances of the finding of a lap seam crack in time to prevent a disastrous explosion were described in the last issue of THE LOCOMOTIVE and well illustrate the really necessary service of the boiler inspector. The lap seam crack is a common cause of failure and many other illustrations can be cited of the very common misunderstanding that exists as regards the use of a boiler in which this weakness exists. Since the publication of the previously mentioned article on this type of defect we have been reminded of several cases which may well be described.

Some years ago a representative of this Company called upon a mill owner in an endeavor to interest him in boiler inspection and insurance, but the owner could see no value in such service. About a month later a crack developed in the longitudinal seam of one of the boilers in this mill and a supposedly experienced mechanic was called in. The crack was drilled at intervals, the holes tapped out, and soft iron plugs inserted to stop the leakage. About a week later the boiler exploded, fatally injuring the fireman and seriously injuring his wife who happened to be there at that time. Had this boiler received the attention of a qualified inspector the dangerous condition could have been pointed out in time to prevent the disaster which occurred.

Mention was made, in the July issue, of repairs to a lap seam crack by autogenous welding. It has been reported to us that a boiler belonging to the Wishkah Shingle Co., State of Washington, was repaired (?) at a lap seam crack by this method and soon afterwards exploded, killing three people. It was said that the boiler was not regularly inspected by a competent person and that the owners had no knowledge of the danger in such a repair.

Many similar instances are recalled of the inevitable sad results which follow improper treatment of a lap seam crack. This type of failure frequently shows itself by a leak extending over a comparatively short distance. The boiler plate may very likely be cracked over a considerable distance but, not being apparent by casual inspection, this is not realized and repairs are confined to the vicinity of the leak. A soft patch is frequently placed over the leak, sometimes temporarily alleviating the leaky condition and at other times accentuating it. Sooner or later, however, the crack extends through the plate at other points and an explosion frequently results.

But the lap seam crack, although a frequent offender, is not the only cause of trouble. Many conditions, dangerous to life and property, are found on the trips of the boiler inspector. Cracks in other parts of a boiler, such as to make it unsafe to use, have often been found. In one such case, cracks twelve and fourteen inches long were found at the bend of the flange on each side of the fire door sheet of a locomotive type boiler and in addition, some smaller cracks in the side sheets, making it necessary to condemn the boiler as unfit for service. A second hand dealer purchased the boiler and sold it to a saw-mill in Virginia. A few days after it was put into service it exploded and two persons were killed. In this case the inspector gave protection to the original owner and also did all he could to safeguard the public by warning the second hand dealer not to sell

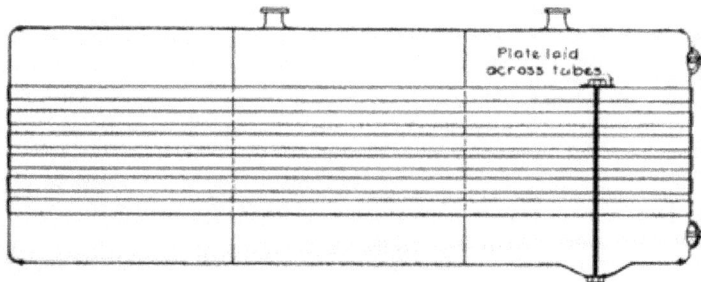

Fig. 1.

the boiler but to cut it up for scrap. The unscrupulous dealer, however, was not so mindful of the safety of others. A similar case was that of a boiler which exploded, some years ago, near Clifton Heights, Pa.

Repairs, that to the inexperienced appear safe, but to the boiler inspector are immediately recognized as dangerous, are frequently found. In one such case a bag had developed in a boiler from overheating which resulted probably from an accumulation of scale. The "repair" made in this case is illustrated in Fig. 1, above. The center of the bag or bulge was tapped and a $\frac{7}{8}$" threaded rod screwed in with the end upset or riveted over. The upper end was supported by a flat iron bar resting on the top row of tubes. The idea apparently was to prevent a complete failure at the bag. However, an excellent place for scale to accumulate was provided by permitting the bulge to remain and the condition presented was rather dangerous. Our inspector had the rod removed. The plate was then heated and the bag driven back, the small hole being plugged with a rivet.

The corrugated furnace in one of two Scotch-marine boilers collapsed and necessitated repairs to that boiler. Thinking it advis-

able to investigate conditions in the other boiler, a hydrostatic test
was applied by our inspector and the furnace showed a decided
tendency to collapse, indicating that the structure was weak for the
pressure desired. It was recommended that a new furnace be installed and assurance was given that this would be done. A few
weeks later the owners requested that we make an examination and
this we proceeded to do. The results were rather remarkable. It
was found that the repair man who was called in to do the work,
after looking over the boiler, and announcing that he could " fix her
as good as new," apparently concluded that a cylindrical furnace,
lying in a horizontal position, could only collapse from the upper
side although subjected to pressure from all sides. He therefore
placed four 1½" rods extending from the upper surface of the
furnace, where the furnace had first shown a tendency to collapse,
to the top of the boiler shell and then remarked to the engineer,
" There she is, she's all yours." Unfortunately for the owner, this
repair could not be approved.

Improper attachment or use of boiler appliances often results
in damage to the boiler even if an explosion does not occur. Such
conditions are often found and corrected by the boiler inspector. As
an example we cite the following: —

One of our inspectors, when calling at the office of a certain
plant, was advised that considerable trouble had been experienced
from leakage at the girth seams of the horizontal water tube boilers
in use in the plant. The superintendent stated that they had recaulked
the seams a number of times but to no avail and the boiler maker
who had done this work had expressed the opinion that they had
received a " bum bunch of boilers." On walking into the boiler room
the first thing that struck the inspector's eye was that the water
glasses were considerably below the drums of the boiler. A comparison of the water level in the gauge glass with the height of the
drums showed that the water columns were set 12" too low so that
when the normal water level was carried in the water column there
was absolutely no water in the drums. This resulted in rapid overheating as soon as the boiler was fired up. When the water column
was correctly placed and the seams again caulked no further trouble
was experienced.

Improper pipe connections are often found. As an example of
this condition, the following experience, related by one of our inspectors, will be illustrative.

"As I was going about my regular work I stopped at a small mine

and became engaged in conversation with the engineer who claimed
he was a licensed man and said that he had installed the boiler plant.
He took great pride in showing me the installation but after looking
it over I was unable to work up the same degree of enthusiasm
that he had and the reason may easily be explained.

" The faulty connection that first struck my eye was that a ½"
pipe had been run from the dome of the boiler to supply an injector
and a feed pump, both of which required a 1" line, and to this pipe
the steam pressure gauge had also been attached. The steam pressure,
when first noted, was 100 pounds. Soon, however, the injector was
started and, because of the pressure drop which then took place in
the ½" pipe, the gauge reading fell back to 60 pounds. The engineer
noticed this and immediately forced his fire a little harder which soon
resulted in the safety valve blowing off although the gauge reading
did not rise any great amount. Seeing this, the engineer remarked,
' Sometimes that safety valve doesn't work just right. I don't know
what gets the matter with it. I've had it off several times and examined it but I can't find anything the matter with it.'

" I stepped outside to look up at the valve which was about twelve
feet above the boiler and while looking at it the blowing stopped.
A few minutes later I walked in and looked at the pressure gauge.
It showed 140 pounds. I immediately started an investigation to
determine the reason for the safety valve remaining closed with this
pressure behind it and, when I climbed up on the boiler, I soon discovered the cause. There I found a gate valve, closed tight, in the
line leading to the safety valve. Apparently the engineer had closed

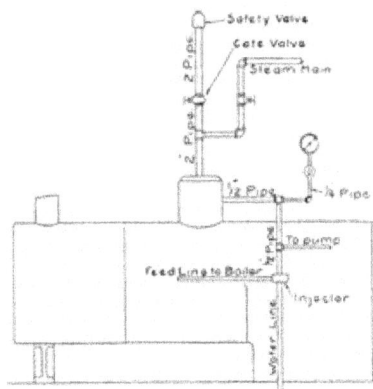

Fig. 2.

it when I stepped outside for just
then I saw him come around to
the front of the boiler, glance at
the pressure gauge, and then
rush up the ladder to open the
gate valve. This I cautioned
him to do slowly as a sudden
release of the pressure on the
boiler might have caused an
explosion. As soon as a few
turns had been given the gate
valve to open it the safety valve
began to blow and continued to
do so until the pressure was
reduced to 100 pounds.

"A sketch of the boiler with its improper pipe connections is given in Fig. 2. In addition to what is shown there I might mention that the blow off line had been piped outside the building where a trap had been formed so that any water accumulating in it would freeze very readily. A man, in blowing down the boiler, would have been placed in a very dangerous position.

"A short time later I had to make an internal inspection of this same boiler and, although the engineer said that he had just had the boiler open the day before and that I could take his word as to its good condition, I found it to be in very poor shape. Although no great amount of scale was present, a considerable amount of corrosion and a number of cracks in different parts of the boiler were found. When this information was given the superintendent he ordered the boiler out of service."

Inspections are of value not only in the detection of dangerous conditions but also in preventing unnecessary expenditures for repairs. In one case a tube ruptured in a vertical water tube boiler and there was also some slight overheating of several other tubes. The coal mining company who owned the boiler engaged a repairman to put the boiler in serviceable condition. This repairman had them place an order for new tubes which were to cost $1200. He expected to almost take apart the entire boiler, the total cost to be about $2,500, and had actually started work. Someone, however, suggested to the coal company that they obtain the services of a boiler inspector and at their request we rushed a man over to the plant. As a result of his inspection the repairs necessary were limited to ten new tubes and some other minor repairs, amounting in all to about $250, or a saving to the owner of over $2000. This was at pre-war prices and the figures would be considerably greater at the present time.

In another somewhat similar instance the expenditure for repairs was reduced from $1100 to about $20. In still another case a boiler had been condemned, by a steam fitter, as unfit for further use but upon examination by a boiler inspector was found to be good for many years more of service.

The boiler inspector also finds a field of service in the help that can often be given a manufacturer in the arrangement of his steam generating equipment and piping. In a certain plant where rendering tanks were used, considerable difficulty had been experienced in obtaining a lard of uniform quality. Changes that were recommended by the inspector resulted in safer working conditions and a more uniform product.

In another striking instance, which seems almost incredible but which, nevertheless, is true, a furniture factory complained of the wastefulness of their boiler equipment from the standpoint of fuel consumption. After an inspection we recommended the installation of a boiler of a different type. Before the change the plant used a car of coal a month in addition to the refuse from the mill. After the change the coal consumption was reduced to a maximum of two cars of coal per year. It does not require much figuring to compute the saving that would be made in a case like this.

Many accounts might be given of the savings resulting from the suggestions of the boiler inspector, such as the covering of steam pipes, use of pumps and exhaust steam heaters in place of injectors, use of a lesser number of boilers to carry the load, and so forth. We shall content ourselves, however, with the relation of one instance which is interesting from more than one point of view.

One of our inspectors, upon visiting a plant, noticed that the exhaust steam drips were being discharged to the sewer. He made the recommendation that this practice be discontinued and, with the proper piping arrangements, the condensate be returned to the boiler so that whatever heat was contained therein might be saved. The next visit of inspection was made by a different man and the engineer of the plant, in speaking to him of the recommendations made by the previous inspector, said, " Now wasn't he a numskull? Why we had already cooked all the steam out of that water."

It would be practically impossible for the owner of a plant to give the time for a personal inspection of his boilers even though he were so inclined and sufficiently experienced in this line of work. No matter how efficient operating engineers may be in running their plant, but few of them have had the specialized training to qualify them as inspectors of boilers. On the other hand the expert boiler inspector examines the boilers for all known causes of trouble and then the owner is handed an unbiased statement of the condition of the equipment. If everything is in good shape the owner has a sense of security from accident, or, if there is trouble in sight, ample time is given to make repairs before the danger point is reached.

A Comparison of the Reciprocating Engine and the Steam Turbine.

MANY persons have had difficulty in understanding the action of the steam in the cylinder of a steam engine and still greater difficulty in obtaining a grasp of the principle of operation of the steam turbine. Many who are well informed as to the pressure and volume changes that take place in the reciprocating engine do not fully comprehend the changes that take place in the steam turbine; and still others, though fully aware of the changes that take place, have had difficulty in appreciating the fact that there is a striking similarity in certain fundamental principles underlying the operation of the reciprocating engine and the turbine. It is the purpose of this discussion to clear up, if possible, some of these obscure points.

We have intimated, only a few lines above, that a certain fundamental similarity exists between two principal types of steam prime movers. In making this statement, we refer to the similarity that exists in the heat changes that take place and which we shall discuss more fully later on. Before proceeding with this discussion, however, it may be well, for our readers who lack the information, to describe the principle of operation of the engine and the turbine. We trust that we will be pardoned if to some, this description seems elementary.

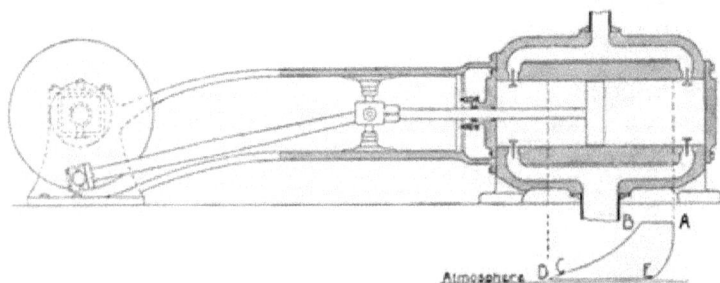

FIG. 1.

The reciprocating steam engine may be illustrated diagrammatically as in Fig. 1. At each end of the cylinder are shown two valves, those at the top being for admission of steam and those at the bottom being for the exit or the exhaust of the used steam. Considering the head end (the right hand end in this sketch) of the cylinder, we have first that the admission valve, operated by the valve gear which is driven by the engine, is opened and steam practically at

boiler pressure is admitted. **The pressure** thus created in the cylinder
drives the piston forward, thereby turning **the** crank shaft. **When
the piston** has moved about one-quarter of its stroke the admission
valve closes. The steam that is thereby entrapped in the cylinder **is
at high pressure and the piston,** therefore, continues its motion, **but
as soon as it** does so the volume occupied by the steam increases and
the pressure begins to fall. Nevertheless, the pressure continues for
some time to be sufficient to drive the piston forward. Under normal
operating conditions the **engine is so** designed that **as soon as the**
pressure of this steam falls to **a point** such that it can no longer be
effective in driving the **piston the exhaust valve at the head end is**
opened and the steam **permitted to escape.** This action, known as
" **release,**" may be considered to take place when the piston has completed about nine-tenths of its forward stroke. The exhaust valve
remains open and the steam escapes from the cylinder during the
remaining one-tenth of the forward stroke, and through the greater
part of the return stroke when the piston goes from the crank end
to the head end of the cylinder. The exhaust valve closes when the
piston has completed about eight-tenths of the return stroke, **and the**
steam remaining in the cylinder is then compressed thereby **reducing
the volume and at** the same time increasing the pressure. **This** action
of compressing the steam assists in giving smooth operation **over
the end or dead** points when the reciprocating parts **change their
direction of motion.** For this reason, the term " **cushioning**" is often
applied to this action of using some of the exhaust steam as a buffer
to take the shock out of the end of the stroke. At or near the end
of the stroke the steam admission **valve opens again and the whole
cycle of events is repeated.** With a **double acting engine there is also**
a similar action taking place in **the crank end of the cylinder.**

The changes in pressure and volume taking place within the
cylinder may be represented on a chart in which horizontal distance
represents volume or length of stroke and vertical distance represents
pressure. Such a chart is shown directly below the cylinder in Fig.
1. Its length has been taken the same as that of the cylinder. The
line AB shows the pressure that exists while the piston is passing
from the end of the stroke (directly **over A**) to the point of cut off
(directly over B). The curved line BC shows that expansion is
taking place or, in other words, that the pressure is falling gradually
with the forward motion of the piston. At the point C the exhaust
valve opens and the pressure within the cylinder then falls from C
to D very rapidly, but not instantaneously, to the pressure in the

exhaust pipe because the steam, most obviously, cannot all get out at once. From D to E the steam is pushed out of the cylinder by the returning piston. The line marked "Atmosphere" is drawn to represent the zero *gauge* pressure of the external air and the line DE is slightly above it because some pressure difference is necessary between the inside and the outside of the cylinder to make the steam flow out. The line EA represents the compression or cushioning effect that takes place at the end of the stroke.

The steam engine indicator is an instrument for securing such a chart in reduced size from an engine. The diagram or chart illustrated in the sketch is, in a way, ideal. Actually the events of cut off, release, compression and admission do not take place with such marked sharpness as is indicated and the actual diagram is somewhat rounded and otherwise changed from the ideal. It also varies greatly in different engines and with the power developed in any one engine.

We have said that there is a striking similarity in the fundamentals of the reciprocating engine and the turbine. This is true from a certain viewpoint and yet, on the other hand, there is a most striking dissimilarity between the two machines. In the turbine, the steam from the boiler is passed through a nozzle or a passage and in so doing is reduced in pressure or expanded. This means that an increase in volume takes place and it also results in giving a high velocity to the steam as it leaves the nozzle. This jet of rapidly moving steam is directed against a number of specially designed vanes, paddles, or buckets fastened on the periphery of a wheel or a drum and by this means rotary motion is obtained. Thus the turbine is essentially a machine that uses the *velocity* of the steam whereas the reciprocating engine depends on the *pressure* of the steam to develop its power. In this respect, therefore, the two machines are widely different. The similarity in fundamental principles is based on other considerations and is a matter we shall now discuss.

Let us suppose that we have a specially constructed boiler, engine and connecting steam pipe, all built with the same cross section as shown in Fig. 2 on the opposite page. Suppose further that steam has been generated at some given pressure, until all the space is filled between the surface of the water in the boiler and piston in the cylinder. We might consider that the space mentioned is filled with several unit volumes of steam in contact with each other, as indicated by the dotted lines in the boiler and steam pipe, and of such a size that each weighs one pound. Now we can cause the piston to move and to increase the volume between it and the position shown in the sketch by pro-

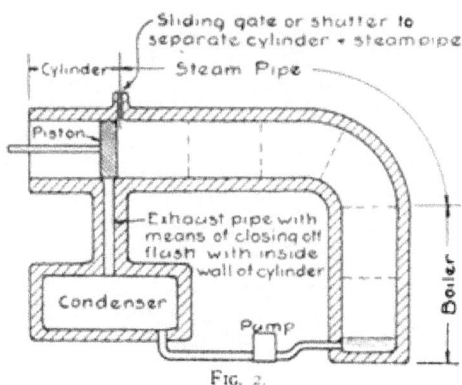

FIG. 2.

ducing or generating another unit volume of steam at the surface of the water in the boiler. To do this, heat, of course, is required.

The changes in volume and pressure that take place in this imaginary engine cylinder can be charted on a diagram similar to that in Fig. 1. This diagram is given in Fig. 3, and the line AB which represents the increase in volume at constant pressure within the cylinder is exactly similar to the line AB of Fig. 1.

Suppose next that the sliding gate or shutter at the head end of the cylinder is closed, thereby preventing the entry of any more steam into the cylinder. The entrapped steam will then expand, driving the piston further along in the cylinder and the pressure-volume change may be represented by the line BC of Fig. 3 which is similar to the line BC of Fig. 1.

In Fig. 1 we had the engine exhausting to the atmosphere which is a space of constant pressure. In the present arrangement we shall substitute a condenser which, being constantly cooled by water at a uniform temperature, reduces the exhaust steam to water and provides a space of constant pressure. The temperature of the cooling

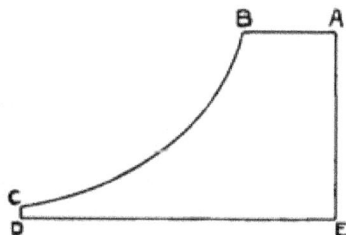

FIG. 3.

water will determine what this pressure will be. Furthermore, in Fig. 1 we had release taking place slightly before the end of the stroke of the piston. In this case, as will be noticed, we do not open the exhaust valve until the completion of the forward piston stroke. The pressure in the cylinder then drops approximately to that in the condenser and, with the return stroke of the piston, we get the line CDE. This line also differs from the corresponding line of Fig. 1 in that we have continued it back to the end of the return stroke.

In this we have differed a little from the actual engine of Fig. 1 in which some of the steam was retained in the cylinder for the cushioning effect. The *weight* of steam so retained, however, is comparatively small and the difference between the real and this imaginary engine is without relative importance.

We now have the piston back to the starting point in contact with the sliding gate and, with the exhaust valve closed, this gate may be opened when the pressure against the piston (we could also say the pressure in the cylinder) is raised. This change can be represented by the line EA.

By means of the pump shown in Fig. 3 the condensed steam may be taken from the condenser and returned to the boiler. In so doing it should be noted that the volume changes but very little, while its pressure is raised from that in the condenser to that in the boiler.

In the above discussion we have been considering one pound of water or steam. During the change from A to B on the chart, we said that to increase the volume we caused one pound of water in the boiler to change to one pound of steam. This involved an increase of volume at constant pressure and theoretically we may say that the line AB is just as representative of the generation of this unit volume of steam in the boiler as it is of the increase of volume in the cylinder. In other words, the pound of steam in the cylinder at one time was a pound of water in the boiler and underwent an increase in volume just as did all the other units of steam between the cylinder and the boiler. Therefore, we may say that the chart in this imaginary engine represents the changes in volume and pressure to which one pound of water is subjected in making the cycle from the boiler through the engine, condenser, pumps and back to the boiler. The line AB represents the increase in volume during the change from water to steam. The line BC represents the expansion of this or a similar pound of steam in the cylinder. The line CDE represents the exhausting of the pound of steam to the condenser and its reduction to one pound of water. Finally, the line EA represents the increase in pressure at practically constant volume, to which the pound of water is subjected in being forced back to the boiler.

Since a steam engine indicator can be applied to a reciprocating engine and can graphically record the changes taking place within the cylinder, it is not so difficult to see the relation between the actual pressure-volume changes in the cylinder and the action of a unit volume of steam or water in passing out from and back to the boiler. It is not possible, however, to take a similar chart from a steam

turbine, yet the vol-
ume and pressure
changes which a
pound of water or
steam undergoes in
the turbine are ex-
actly the same as in
the engine. This we
will attempt to ex-
plain and in doing so,
we will use an imag-
inary arrangement
similar to that of Fig.
3 but adapted for a
steam turbine.

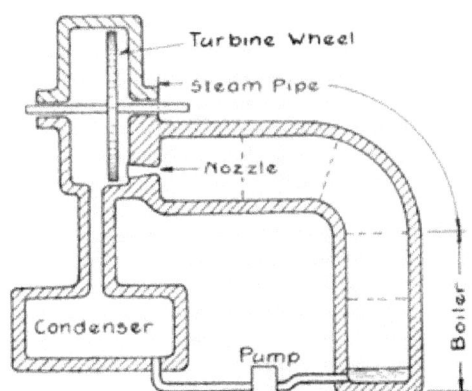

FIG. 4.

This sketch, Fig. 4, shows, diagrammatically, a boiler with a steam
pipe of the same cross section leading to the nozzle of the turbine.
If we are to maintain constant pressure in the boiler while a pound
of steam flows from the pipe into the nozzle of the turbine, we must
change one pound of the water in the boiler into one pound of steam,
thereby pushing along the successive unit volumes of steam and in
turn causing the flow of one unit into the nozzle. Considering the

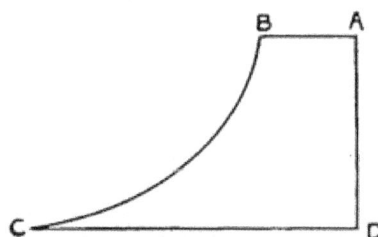

FIG. 5.

chart shown in Fig. 5 as representing the pressure and volume
changes taking place in one
pound of steam or water the line
AB will indicate this first action
of changing from water into
steam at constant pressure.

The turbine differs from the
reciprocating engine in that the
flow is continuous in the turbine

and is intermittent in the engine. In the imaginary engine of Fig. 2
it was proper for us to isolate a unit volume of steam in the cylinder
by the closing of the sliding gate because such action is exactly similar
to the operation of the real engine. With the turbine, however, we
cannot isolate a unit volume. Nevertheless, we can confine our consideration of the pressure and volume changes to one unit, bearing in
mind that every unit that has preceded and everyone that follows after
is acting in the same way.

The pound of steam expands in its flow through the nozzle, the

volume increasing and the pressure falling, so that at the mouth of
the nozzle the pressure has fallen to its lowest value. Such a change
can be represented on our chart by the line BC. It will be noted
in this chart for a turbine that this expansion line continues until
it meets the exhaust line CD in a point and is not chopped off by a
vertical line, such as CD in Fig. 3. It will be remembered that in
the steam engine the expansion was stopped when the *pressure* had
fallen so low that it would no longer be effective in driving the
piston. In the turbine the continuation of the expansion beyond the
point D of Fig. 3 results in increased *velocity* to the steam and
thereby enables the turbine to make use of this " tail end " of the
diagram.

The expansion of the steam in the nozzle increases its velocity
to a high degree and it is this rapidly moving steam impinging on
the blades of the turbine that gives motion to the turbine shaft. In
the type of turbine under consideration, there are no further changes
in pressure or volume while the steam is passing " through the turbine
blades." Power is produced at the turbine shaft by reason of the
reduction in the velocity of the steam and, ideally at least, the steam
leaves the rotating blades at zero velocity. In some types of turbines
the steam passes through several successive sets of nozzles and blades,
the expansion taking place in steps, sometimes wholly in the nozzles
and sometimes partly in the nozzles and partly in the moving blades.
From our present viewpoint, however, the action of expansion is
fundamentally the same as though it took place in one nozzle.

Since there is no further change in pressure and volume while the
steam is acting on the blades of the turbine, it is impossible to chart
this part of the steam flow. After the expansion in the nozzle and
when the steam has been brought to the pressure and volume conditions represented by the point C in Fig. 5, the next change to take
place, from our viewpoint, is the condensation of the steam in the
condenser. This action is exactly the same as in the case of the
steam engine and can be represented by the line CD in Fig. 5. So
also, the return of the condensed steam to the boiler by the feed
pump can be indicated by the vertical line DA.

The two charts shown in Figs. 3 and 5 thus represent the pressure
and volume changes through which a pound of water or steam passes
in its working cycle. In both types of prime mover the increase in
volume represented by the line AB may be thought of as taking place
in the cylinder or nozzle or in the boiler. The expansion of the
steam represented by BC takes place in the cylinder of the engine

or in the nozzle of the turbine. In both machines the reduction of the steam to water and the return of this water to the boiler is accomplished by external means, namely the condenser and the feed pump. Thus we may say that, while the steam engine is a pressure machine and the turbine a velocity machine, the cycle of events through which the steam goes in its passage through either is practically identical

Explosion of an Ammonia Compressor.

AN accident which is of considerable interest and which should serve as a warning to many owners and engineers of ice and cold storage plants using ammonia compressors is one that occurred a short time ago in one of the southern states.

The machine which figured in the accident was a two cylinder single acting compressor, belt driven from a motor. It had been running with apparent ease and smoothness when suddenly the closed crank case was shattered from internal pressure, as illustrated on page 242, the damage being so great that it was necessary to install another complete machine. Fortunately the machine was covered by a Hartford Engine Policy and the owners were promptly indemnified for the property loss they sustained from the accident.

While the breaking of the crank case was clearly the result of over pressure, what gave rise to this condition is not definitely known. The theory has been advanced that ammonia vapor leaking by the pistons had an opportunity to build up to such a degree as to burst the walls of the crank case. It is possible, although not very probable in this incident, that the ammonia fumes became ignited within the crank case thereby causing the explosion. It is known that when ammonia vapor is mixed with a certain proportion of air and raised to a sufficiently high temperature that it will ignite and burn with explosive violence if the mixture is confined. (See THE LOCOMOTIVE, July, 1920, pg. 76.) With the circumstances of this particular case, however, it is difficult to see how the temperature necessary for ignition was produced. On the other hand it may be that the necessary conditions were produced and that this accident did result from the explosive property of ammonia gas. It is also possible that the explosion resulted from the ignition of oil vapor in the crank case although this would also require the presence of a high enough temperature to produce combustion.

In connection with this accident and as further evidence of the inflammability of ammonia we desire to call attention to a disastrous fire which took place at the Syracuse Cold Storage Company of Syracuse, N. Y. on July 13th, 1921. This, it was said, resulted from an ammonia pipe that burst, filling the room with the gas which became ignited by the spark produced when the switch was opened to stop the electric motor driving the machinery. It was reported that five men were injured in fighting the fire and that the property damage amounted to over $50,000.

WRECKED AMMONIA COMPRESSOR.

Explosion of an Open Feed Water Heater.

WHEN is an "open heater" not an "open heater"? The recent explosion of a feed water heater of this type will serve to show that there is a possibility, through faulty pipe connections, for such equipment to become "closed" and capable of doing severe damage.

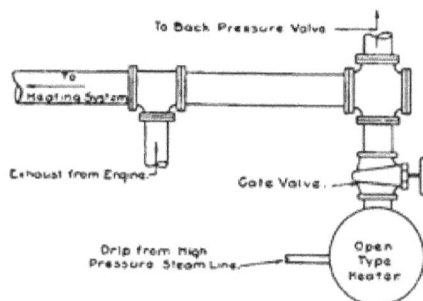

To Back Pressure Valve

To Heating System

Exhaust from Engine.

Gate Valve.

Drip from High Pressure Steam Line.

Open Type Heater

PIPE CONNECTIONS TO HEATER.

A plan of the arrangement of the heater and its piping is given in the accompanying sketch. The engine exhausted into a line which led, in the one direction, to the heating system and, in the other direction, to a back pressure valve and also to the heater, which could be cut off from the exhaust steam supply by a gate valve. The drips from a high pressure steam line were also led into the heater so that some of this heat could be returned to the boiler.

As the heater was in need of cleaning, the gate valve connecting it to the exhaust steam line was closed. The heater was not immediately opened, however, and as it apparently had no vent and leakage existed in the steam traps on the drips of the high pressure steam line, it was not long before a pressure was produced in the heater sufficient to burst it. A view of the wreck is given in the illustration herewith.

Is your "open" heater so installed that it is always *open?* It might be well to have it inspected. Go a step further — have it insured in The Hartford.

WRECK OF AN OPEN FEED WATER HEATER.

The Locomotive

DEVOTED TO POWER PLANT PROTECTION

PUBLISHED QUARTERLY

W. D. HALSEY, Editor.

HARTFORD, OCTOBER, 1921.

SINGLE COPIES *can be obtained free by calling at any of the company's agencies.*
Subscription price 50 cents per year when mailed from this office.
Recent bound volumes one dollar each. Earlier ones two dollars.
Reprinting matter from this paper is permitted if credited to
THE LOCOMOTIVE OF THE HARTFORD STEAM BOILER I. & I. CO.

Obituary.

Francis Burke Allen.

IT is with a sense of deep sorrow that we record the death of Francis Burke Allen, vice-president of this Company, at his home in Hartford, Conn., on July 27th, 1921.

Mr. Allen had been in the service of the Company for forty-nine years, having joined the New York Department in 1872 as special agent. In 1882 he was appointed supervising general agent at the Home Office, in 1888 was elected second vice-president, and on February 9th, 1904 became vice-president of the Company. For many years he directed the affairs of the Inspection Department beside adjusting many claims and to his ability and continued efforts are due largely the development of the inspection service and methods along the lines that have gained so excellent a reputation for The Hartford Steam Boiler Inspection and Insurance Company. During the past few years his failing health confined him to his home and forced him to give up the very active part he had taken in many affairs.

FRANCIS B. ALLEN.

Mr. Allen was born in Baltimore, Md. on June 1st, 1841. He was the son of William Cathers Allen and Louisa Williams Allen. He received a public school education in Baltimore, Philadelphia and Portland, Me. and later served an apprenticeship of four years in the machinist trade. Born of fighting stock — a great grandfather, Edward D. Burke, having served in the Revolutionary War, and one of his grandfathers, Dr. Francis Burke of Washington, D. C., having fought in the War of 1812 — Francis Burke Allen early felt the call of his country and in 1862 entered the United States Navy as assistant engineer with the rank of ensign. In this capacity he saw service during the Civil War on the gunboat Port Royal, which took a prominent part in operations on the James, Appomattox, Chickahominy and Mississippi Rivers and in the battle of Mobile Bay. He was then promoted to the grade of master and was ordered to the ironclad ram Dictator. After a year's assignment to this vessel, service upon which Mr. Allen was able to withstand only by reason of his robust health, he was ordered to the Novelty Iron Works, of New York City, to take part in experiments on the expansion of steam. He later served on the DeSoto, flagship of the West Indies Squadron. In 1868 he resigned from the navy and entered civil life as an engineer in the employ of The Novelty Iron Works remaining with this concern until he entered the service of the Hartford Steam Boiler Inspection and Insurance Company.

Mr. Allen was an intense patriot and this found expression in the active part he took in many patriotic organizations. He was a member of the Grand Army of The Republic, The National Association of Naval Veterans, The Army and Navy Club of Connecticut, and The Military Order of the Loyal Legion and also took an active part in the formation of the Admiral Bunce Section, Navy League of The United States.

In addition to these societies, Mr. Allen was a member of The American Society of Mechanical Engineers and of The American Society of Naval Engineers.

To all who knew him, Mr. Allen endeared himself by his kind and generous spirit. His cheerful personality won him many friends to whom, as to this Company, he was ever loyal. His death has brought to all a feeling of great loss.

Failure of Autogenously Welded Boiler.

THE Hartford Steam Boiler Inspection & Insurance Company has long taken the stand that, except in a few special cases, it could not approve any boiler construction or repair by autogenous welding. That this decision is justified is well illustrated by a recent failure resulting from a repair of this nature in the firebox of a locomotive type boiler.

In this particular case, which was described in *Power*, the boiler was not in service and at the time of the failure was being examined by an inspector who had gone into the firebox. While he was there, he heard a sudden, deafening report and thought at first that a stick of dynamite had exploded nearby. He discovered, however, that one of the side sheets of the firebox had ripped open for a length of about 36 inches, directly through an autogenous weld that had been made in the sheet, the rupture extending about an equal distance on either side of the repair.

The explanation given of the failure was that the sheet was thought to have been overheated when the welding was done so that it became crystallized and was, therefore, under a severe shrinkage strain. It is believed that many other disastrous accidents which have involved welding may have resulted from a similar cause.

The case, while unique, strongly supports our contention that, in cases where the safety of the structure is dependent on the strength of the weld, autogenous welding on pressure vessels is a dangerous practice.

Madman Burglar Steals Steam Boiler.

ORLEANS, FRANCE.— A few nights ago burglars broke into an ironmonger's warehouse and the next morning the only object missing was a large boiler. The police at first believed the theft to be the work of a madman. Later they arrested Mr. Boltier, a wealthy wholesale wine merchant and owner of a chateau and extensive grounds at Cerdon-sur-Loire.

Boltier admitted the theft saying that he had committed it in a moment of weakness, as he had ample means to buy such a boiler had he wished. A number of similar boilers, some of them weighing over a ton, were found in the cellars of his country home.

Boltier's lawyer says he will plead kleptomania.— *Meriden Record.*

[*Editor's Note.— We wouldn't insure these boilers. If M. Boltier had his cellar in the U. S. we wouldn't insure anything in it.*]

During the evening of November 21st, 1920 a heating boiler exploded at the home of Fred Wood, at Grafton, Mass., injuring four persons and causing considerable damage to the building. Mr. Wood had gone to the cellar of his home to remove the drafts from the fire and just as he opened the door of the steam heating boiler the explosion took place, completely wrecking the boiler and literally bombarding Mr. Wood with a volley of iron and hot coals. It was said that the explosion was caused by too great a steam pressure which the safety valve, remaining closed, failed to relieve.

A "low pressure" heating boiler exploded at the Rose Tree Hunt Club near Media, Pa. on March 29, 1921, injuring one person and damaging the club rooms to the extent of about $4,000. The boiler was not provided with a safety valve and the conditions surrounding the accident indicated that the boiler had been fired up while the inlet and outlet valves were closed. The boiler was a small one and a club member remarked after the accident, "If a little thing like that could do this damage I would hate to be near a big one when it goes off on a tear."

A boiler used for hot water supply exploded on March 18th, 1921 at the Hillcrest Apartments, 430 West 116th St., New York, N. Y. The boiler itself was demolished and considerable damage was done to the building. The boiler was one of a battery of two, both of which could be cut off from the water heating system by valves on the inlet and outlet pipes. It is thought that these two valves on the wrecked boiler had not been opened before placing the heater in service and that the safety valve had become inoperative, thus permitting an excessive pressure to build up and cause an explosion.

Boiler Explosions.

(Including Fractures and Ruptures of Pressure Vessels)

MONTH OF DECEMBER, 1920 (Continued)

No.	DAY	NATURE OF ACCIDENT	Kind	Injured	CONCERN	BUSINESS	LOCATION
628	22	Heating boiler exploded		1	Snowden Smith	Residence	Syracuse, N. Y.
629	22	Boiler of locomotive exploded			Santa Fe Railroad	Railroad	Monument, Col.
630	22	Boiler exploded		2	W. F. Hutchins	Saw Mill	Bardstown, Ky.
631	22	Tube ruptured		2	The Texas Co.	Oil Refinery	Norfolk, Va.
632	23	Gauge-glass burst		2	U. S. S. Kennedy	U. S. Navy	Point Loma, Cal.
633	23	Three sections heating boiler cracked			Edwards Hotel	Hotel	Birmingham, Ala
634	23	Two sections heating boiler cracked			St. Lawrence County, N. Y.	Almshouse	Canton, N. Y.
635	23	Two sections heating boiler cracked			S. Laceletti & J. Beally	Hotel	Pleasant Unity, Pa.
636	23	Two sections heating boiler cracked			Borg & Beck Co.	Machine Shop	Moline, Ill.
637	24	Five sections heating boiler cracked			Allen School	School	Canton, O.
638	26	Boiler exploded			Carey Salt Works	Salt Plant	Hutchinson, Kan.
639	26	Eight headers cracked			Lexington Laundry Co.	Laundry	Lexington, Ky.
640	26	Section of heating boiler cracked			Mazer Cigar Mfg. Co.	Cigar Factory	Detroit, Mich.
641	26	Accident to boiler		3	Yellow Cab Co.	Garage	Chicago, Ill.
642	26	Two sections heating boiler cracked			Morris Raphael	Apt. House	N. Britain, Conn.
643	27	Boiler exploded		9	Star Laundry	Laundry	Ottawa, Kans.
644	28	Heating boiler exploded			Traders State Bank	Bank	Glen Elder, Kans.
645	28	Fitting on steam pipe ruptured			Michigan Alkali Co.	Chemical Plant	Wyandotte, Mich.
646	28	Heating boiler exploded			F. J. Fisher	Residence	Detroit, Mich.
647	28	Reducing valve ruptured		2	Illinois Maintenance Co.	Maintenance Co.	Chicago, Ill.
648	29	Boiler exploded. See Locomotive for April, 1921					
649	29	Section of heating boiler cracked			G. H. Tilton & Sons	Hosiery Mfrs.	Savannah, Ga.
650	29	Tube ruptured			Borough of Indiana	School	Indiana, Pa.
651	31	Six sections heating boiler cracked			Morton Salt Co.	Salt Plant	Hutchinson, Kans.
652	31	Boiler exploded		3	G. O. F. Williams	Residence	Gloucester City, N. J.

MONTH OF JANUARY, 1921

No.	DAY	NATURE OF ACCIDENT	Killed	Injured	CONCERN	BUSINESS	LOCATION
1	1	Section of heating boiler cracked			F. W. Woolworth Co.	Store	Watertown, N. Y.
2	2	Accident to steam pipe		1	Milford Elec. & Water Co.	Lt. & Water Plant	Milford, Del.
3	2	Water heating boiler exploded			B. Lebovitz	Residence	Pittsburgh, Pa.
4	3	Heating boiler exploded			Traders State Bank	Bank	Glen Elder, Kan.
5	3	Two sections of heating boiler cracked			Fleischman Co.	Grain Malting	Cincinnati, Ohio
6	3	Twenty-five tubes failed			Charlesbank Homes, Inc.	Apt. House	Boston, Mass.
7	3	Section of heating boiler cracked			Borough of Indiana	School	Indiana, Pa.
8	3	Section of heating boiler cracked			University of Pittsburg	University	Pittsburgh, Pa.
9	4	Hot water heater exploded			T. H. Robinson	Barber Shop	Washburn, Wis.
10	4	Section of heating boiler cracked			J. L. MacIntyre	Garage	Pittsfield, Mass.
11	5	Boiler exploded			Sun Oil Company	Drilling for Oil	Saratoga, Tex.
12	5	Two sections of heating boiler cracked		1	Harmony Cafeteria	Cafeteria	Omaha, Neb.
13	5	Tube ruptured			Montreal Mining Co.	Mining	Montreal, Wis.
14	6	Rendering tank exploded			Baker Slaughter House	Slaughter House	Horton, Kan.
15	6	Throttle valve burst			Shetucket Company	Cotton Mill	Norwich, Conn.
16	7	Vulcanizer exploded		2	Allen Tire & Rubber Co.	Rubber Works	Allentown, Pa.
17	8	Section of heating boiler cracked			County Comm. Hiawatha, Kan.	Court House	Hiawatha, Kan.
18	10	One section of heating boiler cracked			Fleischman Co.	Grain Malting	Cincinnati, Ohio
19	10	Six sections of heating boiler cracked			Fifty Associates	Office Bldg.	Boston, Mass.
20	10	Boiler exploded		1		Threshing Mach.	Hooker, Okla.
21	11	Six sections of heating boiler cracked			Town of Uxbridge	High School	Uxbridge, Mass.
22	11	Tube failure			Cape Girardeau Elec. Co.	Light Plant	Cape Girdeau, M..
23	12	Two sections of heating boiler cracked			Children's Aid Society	Children's Home	New York, N. Y.
24	12	Two sections of heating boiler cracked			Lipman Schurmacher	Garage	New York, N. Y.
25	12	Boiler exploded			Lange Brothers	Cement Works	Champaigne, Ill.
26	13	Heating boiler exploded		1	Kroger Grocery & Baking Co.	Grocery Co.	Detroit, Mich.
27	14	Two sections of heating boiler cracked			E. Gottschalk & Co.	Dry Goods	Fresno, Cal.
28	14	Section of heating boiler cracked			Penn. School No. 6	School	Pittsburgh, Pa.
29	15	Accident to heating boiler		1	Stoeckle Green House	Green House	Watertown, N. Y.
30	6	Tube ruptured			Atlantic City Elec. Co.	Power Station	Atlantic City, N. J.
31	16	Tube ruptured			Leitelt Iron Works	Structural Iron	Gr. Rapids, Mich.
32	17	Section of heating boiler cracked			L. Sinsheimer Est.	Estate	New York, N. Y.
33	17	Six sections of heating boiler cracked			E. M. Glidden	Apt. House	Denver, Col.

No.	Day	Nature of Accident	Owner		Occupation	Location
34	17	Crown sheet pulled down	Massillon Coal Mining Co	1	Coal Mining	Hartford, Ohio
35	17	Two sections of heating boiler cracked	Gotham Can. Co.		Can. Mfgs.	Brooklyn, N. Y.
36	19	Three sections of heating boiler cracked	John R. Thompson		Restaurant	Philadelphia, Pa.
37	20	Heating boiler exploded	Henry Grover		Residence	Two Rivers, Wis.
38	20	Compressed air tank exploded	Deming Filling Station	1		Deming, N. Mex
39	20	Section of heating boiler cracked	Stanley Co. of Amer.			Philadelphia, Pa.
40	22	Boiler ruptured	Bent Wood Farm		Theatre	Preston, Ill.
41	22	Tube ruptured	Los Angeles Gas & Elec. Corp'n		Drilling for Oil	Los Angeles, Cal.
42	22	Two sections of heating boiler cracked	The Garwood Company		Power Station	Brooklyn, N. Y.
43	23	Heating boiler cracked	Superior Garage		Garage	Garwood, N. J.
44	24	Boiler exploded	Jeremiah Sweezy		Wood Saw	Aquebogue, N. Y.
45	25	Two sections of heating boiler cracked	A. B. Wilson & Co.		Garage	Hartford, Conn.
46	25	Four sections of heating boiler cracked	St. Andrews Church		Church	Buffalo, N. Y.
47	25	All sections of heating boiler cracked	Molton Hotel Co.		Hotel	Birmingham, Ala
48	26	Boiler ruptured	Texas & New Orleans R. R. Co.		Creosoting Plant	Houston, Texas
49	26	Plate bulged and ruptured	City of Ada		Light Plant	Ada, Minn.
50	26	Section of heating boiler cracked	New Albany Amuse. Co.		Theatre	Brooklyn, N. Y.
51	26	Section of heating boiler cracked	Fleischman Co.		Grain Malting	Cincinnati, Ohio
52	27	Tube ruptured	Providence Edwards Co.		Gas Plant	Providence, R. I.
53	27	Tube ruptured	Nekoosa Edwards Co.		Paper Mfgrs.	Ft. Edwards, Wis.
54	27	Tube ruptured	Amer. Tel. & Tel. Co.		Telephone Co.	Omaha, Neb
55	27	Heating boiler exploded	Union Mer. & Cooperative Assn.		Mercantile Estab.	Marysville, Kan
56	27	Boiler exploded	J. E. Williamson		Slaughter House	Clendenin, W. V.
57	28	Boiler bagged and fractured	United Fuel Gas Co.	1	Gasoline Plant	Clinton, Ill.
58	28	Boiler exploded	National Drug Co.		Drug Mfgrs.	Montreal, Canada
59	28	Heating boiler exploded	Poulter Apartments		Apt. House	Ardmore, Okla.
60	28	Vulcanizer exploded	Allen Tire & Rubber Co.		Rubber Works	Allentown, Pa
61	29	Two boilers bagged and fractured	United Fuel Gas Co.	5	Gasoline Plant	Clendenin, W. V.
62	29	Boiler of locomotive exploded	B. & O. R. R.		Railroad	Littleton, W. Va
63	31	Heating boiler exploded	C. A. Allen	4	Residence	Hoopeston, Ill

MONTH OF FEBRUARY, 1921

No.	Day	Nature of Accident	Owner		Occupation	Location
64	1	Rupture of valve and fittings	Galveston Ice & Cold Storage Co.	1	Ice & Cold Storage	Galveston, Tex.
65	1	Accident to blow-off pipe	Manchester Laundry Co.	1	Laundry	Philadelphia, Pa.
66	1	Boiler exploded			Threshing Mach.	Liberal, Kan
67	2	Section of heating boiler cracked	G. L. Morganthau	1	Taft Bldg	New York, N. Y.
68	2	Six sections of heating boiler cracked	C. M. Wortman		Hotel	Indiana, Pa
69	2	Boiler exploded		1	Sawmill	Lake Odessa, Mich.

MONTH OF FEBRUARY, 1921 (Continued

No.	DAY	Killed	Injured	NATURE OF ACCIDENT	CONCERN	BUSINESS	LOCATION
70	3			Six sections of heating boiler cracked	A. Kohn & D. Baird	Commercial Bldg.	Louisville, Ky.
71	3		1	Tube ruptured	Texas Power & Light Co.	Power Station	Palestine, Tex.
72	4			Section of heating boiler cracked	Fleischman Malting Co.	Grain Malting	Buffalo, N. Y.
73	4			Section of heating boiler cracked	J. M. & L. Keller	Club Rooms	Brazil, Ind.
74	4			Eleven headers cracked	Congoleum Co.	Floor Coverings	Marcus Hook, Pa.
75	5			Two sections of heating boiler and water front cracked	The Carroll Thomson Co.	Garage	Columbus, Ohio
76	5	1		Boiler exploded		Drilling for Oil	Breckenridge, Kan.
77	6			Header cracked	National Malleable Castings Co.	Wagon Mfg.	Cleveland, Ohio
78	7			Crack in fire sheet	Abingdon Wagon Co.	Malleable	Abingdon, Ill.
79	7			Tube failure	International Paper Co.	Paper Mill	Glen Park, N. Y.
80	8			Crown Sheet collapsed	Pagossa Lumber Co.	Planing Mill	Dulce, New Mex.
81	9			Vulcanizer exploded	Standard Tire & Rubber Co.	Rubber Works	Boston, Mass.
82	9			Vulcanizer exploded	Herman Weddige	Tire Repair Shop	Philadelphia, Pa.
83	10			Tube ruptured	S. H. Greene & Sons	Print Works and	Riverpoint, R. I.
84	13			Section of heating boiler cracked	H. H. Knoblock	Hotel	Fresno, Cal.
85	14			Mud drum fractured	Belton Cotton Mills	Cotton Mill	Belton, S. C.
86	14			Section of heating boiler cracked	S. J. Cordner & Co.	Oil Depot	W. Spg'fld. Mass.
87	14			Section of heating boiler cracked	Fleischman Malting Co.	Grain Malting Co.	Buffalo, N. Y.
88	14			Section of heating boiler cracked	Consol Amusement Enterprise	Theatre	New York, N. Y.
89	15			Section of heating boiler cracked	Central Auto & Supply Co.	Auto Supplies	Jackson, Mich.
90	16			Failure of main steam pipe	Western Compress & Storage Co.	Cotton Compress	Abilene, Texas
91	16			Three sections of heating boiler cracked	Bd. of Ed., Norman, Okla.	School	Norman, Okla.
92	17			Section of heating boiler cracked	Hartford Realty Co.	Apt. House	Hartford, Conn.
93	19			Section of heating boiler cracked	Harney Realty Co.	Hotel	Omaha, Neb.
94	19			Two tubes cracked in heating boiler	The Green Realty Co.	Hotel	Danbury, Conn.
95	19			Section of heating boiler cracked	George Schleicher	Residence	New York, N. Y.
96	20			Failure of mud drum	Lit Brothers	Dept. Store	Philadelphia, Pa.
97	20			Rupture of shell	Gloucester Ice Mfg. Co.	Ice Mfg.	Gloucester, N. Y.
98	20			Three sections of heating boiler cracked	Mayer & Schneider Enterprise	Theatre	New York, N. Y.
99	21			Two sections of heating boiler cracked	Children's Aid Society	Children's Home	New York, N. Y.
100	21			Section of heating boiler cracked	City of Altoona	City Hall	Altoona, Pa.
101	24	4	2	Boiler of locomotive exploded	Lehigh Valley R. R.	Railroad	Jersey City, N. J.

No.	Date	Nature of Accident	Killed	Owner	Type	Location
102	25	Section of heating boiler cracked		J. M. Keller	Mercantile	Brazil, Ind.
103	25	Heating boiler exploded		Smithwood Grammar School	School	Knoxville, Tenn.
104	27	Boiler exploded		St. John's Orphanage	Orphanage	Philadelphia, Pa.
105	27	Three sections of heating boiler cracked		M. C. Barrett	Garage	Springfield, Mass
106	28	Two sections of heating boiler cracked		B. & R. Gross	Apt. House	Newark, Del.
107	28	Tube collapsed		Curtis & Brother	Paper Mfg.	Hartford, Conn.
108	28	Shut-off valve broke		United Laundries Co.	Laundry	Cambridge, Mass.
109	28	Section of heating boiler cracked		Fall River Daily Herald Pub. Co.	Publishing Co.	Fall River, Mass.
110	28	Boiler exploded	1	Missouri Pacific R. R.	Railroad	Van Buren, Ark.

MONTH OF MARCH, 1921

No.	Date	Nature of Accident	Killed	Owner	Type	Location
111	1	Section of heating boiler cracked		Samuel Mathis	Apt. House	Boston, Mass.
112	1	Section of heating boiler cracked		Consol. Realty & Theatre Corp.	Theatre	Evansville, Ind.
113	2	Tube ruptured	2	American Stove Co.	Stove Works	Cleveland, Ohio
114	3	Boiler exploded	1	James Shoemaker Farm	Sawmill	Benton, Ill.
115	4	Failure of blow-off fitting		Whittle Trunk & Bag Co.	Trunk Mfg.	Knoxville, Tenn.
116	4	Boiler bulged and ruptured		Germaine Bros. Co.	Piano Mfg.	Saginaw, Mich.
117	5	Heating boiler exploded		St. Paul's Lutheran Church	Church	Platteville, Wis.
118	5	Heating boiler exploded		Belvedere Apts.	Apt. House	Louisville, Ky.
119	7	Section of heating boiler cracked		T. E. Beaniff	Apt. House	Okla. City, Okla.
120	10	Section of heating boiler cracked		North Congregational Society	Church	Haverill, Mass.
121	10	Boiler of locomotive exploded	2	B. & O. R. R.	Railroad	Barkshak, Md.
122	11	Bulged and cracked fire sheets		Mutual Ice & Cold Storage Co.	Ice Plant	Topeka, Kan.
123	11	Two headers fractured		Nat. Soldiers Home	Soldiers' Home	Hampton, Va.
124	12	Pipe failure		A. L. Aiken	Hospital	New Haven, Conn.
125	12	Failure of three headers	2	City of Baltimore	Rubber Goods	Baltimore, Md.
126	13	Accident to blow-off pipe Boiler No. 2		Northern Pipe Line Co.	Pumping Sta.	Leesburg, Pa.
127	13	Accident to blow-off pipe Boiler No. 1		Northern Pipe Line Co.	Pumping Sta.	Leesburg, Pa.
128	15	Three sections of heating boiler cracked		Pittsburg Bd. of Ed.	School	Pittsburgh, Pa.
129	15	Section of heating boiler cracked		Ironwood Amusement Co.	Theatre	Ironwood, Mich.
130	15	Three sections of heating boiler cracked		Underwriter Bldg. Co.	Office Bldg.	New York, N. Y.
131	15	Section of heating boiler cracked		U. M. Ryan & Co.	Stores & Sales Rms.	Boston, Mass.
132	16	Failure of fitting on blow-off pipe		Nestles Food Co.	Creameries	Menomonee, Wis.
133	17	Failure of pipe fitting		The Frost Mfg. Co.	Machine Works	Galesburg, Ill.
134	17	Header cracked		Pittsburg Plate Glass Co.	Mfgs. Plate Glass	Ford City, Pa.
135	18	Tube failure		South Acton Woolen Co.	Woolen Mills	So. Acton, Mass.
136	18	Heating boiler exploded	1	Hillcrest Apts.	Apt. Home	New York, N. Y.

The Hartford Steam Boiler Inspection and Insurance Company

ABSTRACT OF STATEMENT, JANUARY 1, 1921

Capital Stock, . . $2,000,000.00

ASSETS.

Cash in offices and banks	$366,891.88
Real Estate	90,000.00
Mortgage and collateral loans	1,543,250.00
Bonds and stocks	6,188,435.00
Premiums in course of collection	728,199.44
Interest accrued	116,654.78
Total assets	9,033,431.10

LIABILITIES.

Reserve for unearned premiums		$4,512,194.11
Reserve for losses		205,160.80
Reserve for taxes and other contingencies		388,958.85
Capital stock	$2,000,000.00	
Surplus over all liabilities	1,927,117.34	

Surplus to Policy-holders , . $3,927,117.34

Total liabilities	$9,033,431.10

CHARLES S. BLAKE, President.
FRANCIS B. ALLEN, Vice-President.　　　W. R. C. CORSON, Secretary
L. F. MIDDLEBROOK, Assistant Secretary.
E. SIDNEY BERRY, Assistant Secretary.
S. F. JETER, Chief Engineer.
H. E. DART, Supt. Engineering Dept.
F. M. FITCH, Auditor.
J. J. GRAHAM, Supt. of Agencies.

BOARD OF DIRECTORS

ATWOOD COLLINS, President,
Security Trust Co., Hartford, Conn.

LUCIUS F. ROBINSON, Attorney,
Hartford, Conn.

JOHN O. ENDERS, President,
United States Bank, Hartford, Conn.

MORGAN B. BRAINARD,
Vice-Pres. and Treasurer, Ætna Life
Insurance Co., Hartford, Conn.

CHARLES P. COOLEY, President,
Society for Savings, Hartford, Conn.

FRANCIS T. MAXWELL, President,
The Hockanum Mills Company, Rockville, Conn.

HORACE B. CHENEY, Cheney Brothers,
Silk Manufacturers, South Manchester,
Conn.

D. NEWTON BARNEY, Treasurer, The
Hartford Electric Light Co., Hartford,
Conn.

DR. GEORGE C. F. WILLIAMS, President and Treasurer, The Capewell
Horse Nail Co., Hartford, Conn.

JOSEPH R. ENSIGN, President, The
Ensign-Bickford Co., Simsbury, Conn.

EDWARD MILLIGAN, President,
The Phœnix Insurance Co., Hartford,
Conn.

MORGAN G. BULKELEY, JR.,
Ass't Treas., Ætna Life Ins. Co.,
Hartford, Conn.

CHARLES S. BLAKE, President,
The Hartford Steam Boiler Inspection
and Insurance Co.

WM. R. C. CORSON, Secretary,
The Hartford Steam Boiler Inspection
and Insurance Company.

Incorporated 1866.

Charter Perpetual.

INSURES AGAINST LOSS FROM DAMAGE TO PROPERTY AND PERSONS, DUE TO BOILER OR FLYWHEEL EXPLOSIONS AND ENGINE BREAKAGE

Department.	Representatives.
ATLANTA, Ga.,	W. M. Francis, Manager.
1103-1106 Atlanta Trust Bldg.	C. R. Summers, Chief Inspector.
BALTIMORE, Md.,	Lawford & McKim, General Agents.
13-14-15 Abell Bldg.	James G. Reid, Chief Inspector.
BOSTON, Mass.,	Ward I. Cornell, Manager.
4 Liberty Sq., Cor. Water St.	Charles D. Noyes, Chief Inspector.
BRIDGEPORT, CT.,	W. G. Lineburgh & Son, General Agents.
404-405 City Savings Bank Bldg.	E. Mason Parry, Chief Inspector.
CHICAGO, Ill.,	J. F. Criswell, Manager.
209 West Jackson B'v'l'd .	P. M. Murray, Ass't Manager.
	J. P. Morrison, Chief Inspector.
	J. T. Coleman, Ass't Chief Inspector.
	C. W. Zimmer, Ass't Chief Inspector.
CINCINNATI, Ohio,	W. E. Gleason, Manager.
First National Bank Bldg.	Walter Gerner, Chief Inspector.
CLEVELAND, Ohio,	H. A. Baumhart, Manager.
Leader Bldg.	L. T. Gregg, Chief Inspector.
DENVER, Colo.,	J. H. Chesnutt,
916-918 Gas & Electric Bldg.	Manager and Chief Inspector.
HARTFORD, Conn.,	F. H. Kenyon, General Agent.
56 Prospect St.	E. Mason Parry, Chief Inspector.
NEW ORLEANS, La.,	R. T. Burwell, Mgr. and Chief Inspector.
308 Canal Bank Bldg	E. Unsworth, Ass't Chief Inspector.
NEW YORK, N. Y.,	C. C. Gardiner, Manager.
100 William St.	Joseph H. McNeill, Chief Inspector.
	A. E. Bonnett, Ass't Chief Inspector.
PHILADELPHIA, Pa.,	A. S. Wickham, Manager.
142 South Fourth St.	Wm. J. Farran, Consulting Engineer.
	S. B. Adams, Chief Inspector.
PITTSBURGH, Pa.,	Geo. S. Reynolds, Manager.
1807-8-9-10 Arrott Bldg.	J. A. Snyder, Chief Inspector.
PORTLAND, Ore.,	McCargar, Bates & Lively,
306 Yeon Bldg.	General Agents.
	C. B. Paddock, Chief Inspector.
SAN FRANCISCO, Cal.,	H. R. Mann & Co., General Agents
339-341 Sansome St.	J. B. Warner, Chief Inspector.
ST. LOUIS, Mo.,	C. D. Ashcroft, Manager.
319 North Fourth St.	Eugene Webb, Chief Inspector.
TORONTO, Canada,	H. N. Roberts, President, The Boiler Inspection and Insurance Company of Canada.
Continental Life Bldg.	